Introduction to the Basic
Concepts of Modern Physics

D1195651

WITHDRAWN

Carlo M. Becchi, Massimo D'Elia

Introduction to the Basic Concepts of Modern Physics

Springer

CARLO M. BECCHI
MASSIMO D'ELIA
Istituto Nazionale di Fisica Nucleare - Sezione di Genova

Library of Congress Control Number: 2007927251

ISBN 978-88-470-0606-5 Springer Milan Berlin Heidelberg New York

Springer is a part of Springer Science+Business Media
springer.com

© Springer-Verlag Italia 2007

Printed in Italy

Cover design: Simona Colombo, Milano
Typeset by the authors using a Springer Macro package
Printing and binding: Grafiche Porpora, Segrate, Milano

Springer-Verlag Italia
Via Decembrio, 28
20137 Milano

Printed on acid-free paper

Preface

During the last years of the Nineteenth Century, the development of new techniques and the refinement of measuring apparatuses provided an abundance of new data whose interpretation implied deep changes in the formulation of physical laws and in the development of new phenomenology.

Several experimental results lead to the birth of the *new physics*. A brief list of the most important experiments must contain those performed by H. Hertz about the photoelectric effect, the measurement of the distribution in frequency of the radiation emitted by an ideal oven (the so-called *black body* radiation), the measurement of specific heats at low temperatures, which showed violations of the Dulong–Petit law and contradicted the general applicability of the equi-partition of energy. Furthermore we have to mention the discovery of the electron by J. J. Thomson in 1897, A. Michelson and E. Morley's experiments in 1887, showing that the speed of light is independent of the reference frame, and the detection of line spectra in atomic radiation.

From a theoretical point of view, one of the main themes pushing for new physics was the failure in identifying the ether, i.e. the medium propagating electromagnetic waves, and the consequent Einstein–Lorentz interpretation of the Galilean *relativity principle*, which states the equivalence among all reference frames having a linear uniform motion with respect to fixed stars.

In the light of the electromagnetic interpretation of radiation, of the discovery of the electron and of Rutherford's studies about atomic structure, the anomaly in black body radiation and the particular line structure of atomic spectra lead to the formulation of *quantum theory*, to the birth of *atomic physics* and, strictly related to that, to the quantum formulation of the *statistical theory of matter*.

Modern Physics, which is the subject of this notes, is well distinct from *Classical Physics*, developed during the XIX century, and from *Contemporary Physics*, which was started during the Thirties (of XX century) and deals with the nature of *Fundamental Interactions* and with the physics of matter under extreme conditions. The aim of this introduction to Modern Physics is that of presenting a quantitative, even if necessarily also synthetic and schematic, account of the main features of *Special Relativity*, of *Quantum Physics* and of its application to the *Statistical Theory of Matter*. In usual textbooks these three subjects are presented together only at an introductory and descriptive level, while analytic presentations can be found in distinct volumes, also in view of examining quite complex technical aspects. This state of things can be problematic from the educational point of view.

Indeed, while the need for presenting the three topics together clearly follows from their strict interrelations (think for instance of the role played by special relativity in the hypothesis of de Broglie's waves or of that of statistical physics in the hypothesis of energy quantization), it is also clear that this unitary presentation must necessarily be supplied with enough analytic tools so as to allow a full understanding of the contents and of the consequences of the new theories.

On the other hand, since the present text is aimed to be introductory, the obvious constraints on its length and on its prerequisites must be properly taken into account: it is not possible to write an introductory encyclopaedia. That imposes a selection of the topics which are most qualified from the point of view of the physical content/mathematical formalism ratio.

In the context of special relativity we have given up presenting the covariant formulation of electrodynamics, limiting therefore ourselves to justifying the conservation of energy and momentum and to developing relativistic kinematics with its quite relevant physical consequences. A mathematical discussion about quadrivectors has been confined to a short appendix.

Regarding Schrödinger quantum mechanics, after presenting with some care the origin of the wave equation and the nature of the wave function together with its main implications, like *Heisenberg's Uncertainty Principle*, we have emphasized its qualitative consequences on energy levels, giving up a detailed discussion of atomic spectra beyond the simple Bohr model. Therefore the main analysis has been limited to one-dimensional problems, where we have examined the origin of discrete energy levels and of band spectra as well as the tunnel effect. Extensions to more than one dimension have been limited to very simple cases in which the Schrödinger equation is easily separable, like the three-dimensional harmonic oscillator and the cubic well with completely reflecting walls, which are however among the most useful systems for their applications to statistical physics. In a brief appendix we have sketched the main lines leading to the solution of the three-dimensional motion in a central potential, hence in particular of the hydrogen atom spectrum.

Going to the last subject, which we have discussed, as usual, on the basis of Gibbs' construction of the statistical ensemble and of the related distribution, we have chosen to consider those cases which are more meaningful from the point of view of quantum effects, like degenerate gasses, focusing in particular on distribution laws and on the equation of state, confining the presentation of entropy to a brief appendix.

In order to accomplish the aim of writing a text which is introductory and analytic at the same time, the inclusion of significant collections of problems associated with each chapter has been essential. We have possibly tried to avoid mixing problems with text complements: while moving some relevant topics to the exercise collection may be tempting in order to streamline the general presentation, it has the bad consequence of leading to excessively long exercises which dissuade the average student from trying to give an answer before looking at the suggested solution scheme. On the other hand, we have tried to limit the number of those (however necessary) exercises involving a mere analysis of the order of magnitudes of the physical effects under consideration. The resulting picture, regarding problems, should consist of a sufficiently wide series of applications of the theory, being simple but technically non-trivial at the same time: we hope that the reader will feel that this result has been achieved.

Going to the chapter organization, the one about *Special Relativity* is divided in two sections, dealing respectively with Lorentz transformations and with relativistic kinematics. The chapter on *Wave Mechanics* is made up of eight sections, going from an analysis of the photo-electric effect to the Schrödinger equation and from the potential barrier to the analysis of band spectra. Finally, the chapter on the *Statistical Theory of Matter* includes a first part dedicated to Gibbs distribution and to the equation of state, and a second part dedicated to the Grand Canonical distribution and to perfect quantum gasses.

Genova,   *Carlo Maria Becchi*
April 2007   *Massimo D'Elia*

Suggestion for Introductory Reading

- K. Krane: *Modern Physics*, 2nd edn (John Wiley, New York 1996)

Physical Constants

- speed of light in vacuum: $c = 2.998 \ 10^8$ m/s

- Planck's constant: $h = 6.626 \ 10^{-34}$ J s $= 4.136 \ 10^{-15}$ eV s

- $\hbar = 1.055 \ 10^{-34}$ J s $= 6.582 \ 10^{-16}$ eV s

- Boltzmann's constant : $k = 1.381 \ 10^{-23}$ J/°K $= 8.617 \ 10^{-5}$ eV/°K

- electron charge magnitude: $e = 1.602 \ 10^{-19}$ C

- electron mass: $m_e = 9.109 \ 10^{-31}$ Kg $= 0.5110$ MeV/c^2

- proton mass: $m_p = 1.673 \ 10^{-27}$ Kg $= 0.9383$ GeV/c^2

- permittivity of free space: $\epsilon_0 = 8.854 \ 10^{-12}$ F / m

Contents

1

Introduction to Special Relativity

Maxwell equations in vacuum space describe the propagation of electromagnetic signals with speed $c \equiv 1/\sqrt{\mu_0\epsilon_0}$. Since, according to Galilean relativity principle, velocities must be added like vectors when going from one inertial reference frame to another, the vector corresponding to the velocity of a luminous signal in one inertial reference frame O can be added to the velocity of O with respect to a new inertial frame O' to obtain the velocity of the luminous signal as measured in O'. For a generic value of the relative velocity, the speed of the signal in O' will be different, implying that, if Maxwell equations are valid in O, they are not valid in a generic inertial reference frame O'.

In the Nineteenth Century the most natural solution to this paradox seemed that based, in analogy with the propagation of elastic waves, on the assumption that electromagnetic waves correspond to deformations of an extremely rigid and rare medium, which was named *ether*. However that led to the problem of finding the reference system at rest with ether.

Taking into account that Earth rotates along its orbit with a velocity which is about 10^{-4} times the speed of light, an experiment able to reveal the possible change of velocity of the Earth with respect to the ether in two different periods of the year would require a precision of at least one part over ten thousand. We will show how A. Michelson and E. Morley were able to reach that precision by using light interference.

Another aspect of the same problem comes out when considering the force exchanged between two charged particles at rest with respect to each other. From the point of view of an observer at rest with the particles, the force is given by Coulomb law, which is repulsive it the charges have equal sign. An observer in a moving reference frame must instead also consider the magnetic field produced by each particle, which acts on moving electric charges according to Lorentz's force law. If the velocity of the particles is orthogonal to their relative distance, it can be easily checked that the Lorentz force is opposite to the Coulomb one, thus reducing the electrostatic force by a factor $(1 - v^2/c^2)$. Even if small, the difference leads to different accelerations in the two reference frames, in contrast with Galilean relativity principle. According to this analysis Coulomb law should be valid in no inertial reference system but that at rest with ether. However in this case violations are not easily detectable: for instance, in the case of two electrons accelerated through a potential gap equal to 10^4 V, one would need a precision of the order of $v^2/c^2 \simeq 4\ 10^{-4}$ in order to reveal the effect, and such precisions are not easily attained in the measurement of a force. For this reason it was much more convenient to measure the motion of Earth with respect to the ether by studying interference effects related to variations in the speed of light.

1.1 Michelson–Morley Experiment and Lorentz Transformations

The experimental analysis was done by Michelson and Morley who used a two-arm interferometer similar to what reported in Fig. 1.1. The light source L generates a beam which is split into two parts by a half-silvered mirror S. The two beams travel up to the end of the arms 1 and 2 of the interferometer, where they are reflected back to S: there they recombine and interfere along the tract connecting to the observer in O. The observer detects the phase shift, which can be easily shown to be proportional to the difference ΔT between the times needed by the two beams to go along their paths: if the two arms have the same length l and light moves with the same velocity c along the two directions, then $\Delta T = 0$ and constructive interference is observed in O.

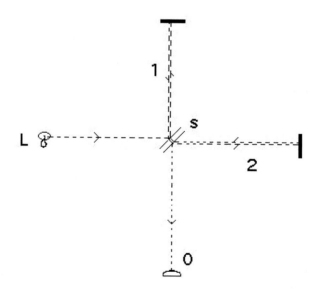

Fig. 1.1. A sketch of Michelson-Morley interferometer

If however the interferometer is moving with respect to ether with a velocity v, which we assume for simplicity to be parallel to the second arm, then the path of the first beam will be seen from the reference system of the ether as reported in the nearby figure and the time T needed to make the path will be given by Pitagora's theorem:

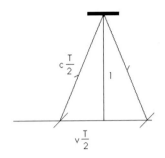

$$c^2\, T^2 = v^2\, T^2 + 4\, l^2 \qquad (1.1)$$

from which we infer

$$T = \frac{2\, l/c}{\sqrt{1 - v^2/c^2}} \; . \qquad (1.2)$$

If we instead consider the second beam, we have a time t_1 needed to make half-path and a time t_2 to go back, which are given respectively by

$$t_1 = \frac{l}{c - v} \; , \qquad\qquad t_2 = \frac{l}{c + v} \qquad (1.3)$$

so that the total time needed by the second beam is

$$T' = t_1 + t_2 = \frac{2\, l/c}{1 - v^2/c^2} = \frac{T}{\sqrt{1 - v^2/c^2}} \qquad (1.4)$$

and for small values of v/c one has

$$\Delta T \equiv T' - T \simeq \frac{T\, v^2}{2\, c^2} \simeq \frac{l\, v^2}{c^3} \; ; \qquad (1.5)$$

this result shows that the experimental apparatus is in principle able to reveal the motion of the laboratory with respect to ether.

If we assume to be able to reveal time differences δT as small as $1/20$ of the typical oscillation period of visible light (hence phase differences as small as $2\pi/20$), i.e. $\delta T \sim 5\ 10^{-17}$ s, and we take $l = 2$ m, so that $l/c \simeq 0.6\ 10^{-8}$ s, we obtain a sensitivity $\delta v/c = \sqrt{\delta T\, c/l} \sim 10^{-4}$, showing that we are able to reveal velocities with respect to ether as small as $3\ 10^4$ m/s, which roughly corresponds to the orbital speed of Earth. If we compare the outcome of two such experiments separated by an interval of 6 months, corresponding to Earth velocities differing by approximately 10^5 m/s, we should be able to reveal the motion of Earth with respect to ether. The experiment, repeated in several different times of the year, clearly showed, together with other complementary observations, that ether does not exist.

Starting from that observation, Einstein deduced that Galilean transformation laws between inertial reference frames:

$$t' = t \, ,$$

$$x' = x - vt \, , \tag{1.6}$$

are inadequate and must be replaced with new linear transformation laws maintaining the speed of light invariant from one reference system to the other, i.e. they must transform the time-line $x = ct$, describing the motion of a luminous signal emitted in the origin at time $t = 0$, into $x' = ct'$, assuming the origins of the reference frames of the two observers coincide at time $t = t' = 0$. Linearity must be maintained in order that motions which are uniform in one reference frame stay like that in all other inertial reference frames.

In order to deduce the new transformation laws, let us impose that the origin of the new reference frame O' appear to be moving with respect to O with velocity v

$$x' = A(x - v \, t) \tag{1.7}$$

where A is some constant to be determined. We can also write

$$x = A(x' + v \, t') \tag{1.8}$$

since transformation laws must be symmetric under the substitution $x \leftrightarrow x'$, $t \leftrightarrow t'$ and $v \leftrightarrow -v$. Combining (1.7) and (1.8) we obtain

$$t' = \frac{x}{A \, v} - \frac{x'}{v} = \frac{x}{A \, v} - \frac{A \, x}{v} + A \, t = A \left(t - \frac{x}{v}(1 - \frac{1}{A^2}) \right) \tag{1.9}$$

from which we see that A must be positive in order that the arrow of time be the same for the two observers.

Let us now apply our results to the motion of a luminous signal, i.e. let us take $x = ct$ and impose that $x' = ct'$. Combining (1.7) and (1.9) we get

$$x' = c \, A \, t \left(1 - \frac{v}{c} \right) \, , \qquad t' = A \, t \left(1 - \frac{c}{v}(1 - \frac{1}{A^2}) \right) \tag{1.10}$$

hence, imposing $x' = ct'$, we obtain

$$\left(1 - \frac{c}{v}(1 - \frac{1}{A^2}) \right) = \left(1 - \frac{v}{c} \right) \tag{1.11}$$

from which it easily follows that

$$A = \frac{1}{\sqrt{1 - v^2/c^2}} \, . \tag{1.12}$$

We can finally write

$$x' = \frac{1}{\sqrt{1 - v^2/c^2}}(x - vt) \, , \qquad t' = \frac{1}{\sqrt{1 - v^2/c^2}} \left(t - \frac{v}{c^2}x \right) \, , \tag{1.13}$$

while orthogonal coordinates are left invariant

$$y' = y, \qquad z' = z, \qquad (1.14)$$

since this is the only possibility compatible with linearity and symmetry under reversal of v, i.e. under exchange of O and O'. These are Lorentz's transformation laws, which can be easily inverted by simply changing the sign of the relative velocity:

$$x = \frac{1}{\sqrt{1 - v^2/c^2}}(x' + vt') , \quad t = \frac{1}{\sqrt{1 - v^2/c^2}} \left(t' + \frac{v}{c^2} x' \right) , \qquad (1.15)$$

Replacing in (1.13) t by $x_0 = ct$ and setting $\sinh \chi \equiv v/\sqrt{c^2 - v^2}$, we obtain:

$$x' = \cosh \chi \, x - \sinh \chi \, x_0 , \quad x_0' = \cosh \chi \, x_0 - \sinh \chi \, x . \qquad (1.16)$$

It clearly appears that previous equations are analogous to two-dimensional rotations, $x' = \cos \theta \, x - \sin \theta \, y$, and $y' = \cos \theta \, y + \sin \theta \, x$, with trigonometric functions replaced by hyperbolic functions. However, while rotations keep $x^2 + y^2$ invariant, equations (1.16) keep invariant the quantity $x^2 - x_0^2$, indeed

$$x'^2 - x_0'^2 = (\cosh \chi \, x - \sinh \chi \, x_0)^2 - (\cosh \chi \, x_0 - \sinh \chi \, x)^2 = x^2 - x_0^2 . \quad (1.17)$$

That suggests to think of Lorentz transformations as generalized "rotations" in space and time.

The three spatial coordinates (x, y, x) plus the time coordinate ct of any event in space-time can then be considered as the components of a *quadrivector*. The invariant length of the quadrivector is $x^2 + y^2 + z^2 - c^2 t^2$, which is analogous to usual length in space but can also assume negative values. Given two quadrivectors, (x_1, y_1, z_1, ct_1) and (x_2, y_2, z_2, ct_2), it is also possible to define their *scalar product* $x_1 x_2 + y_1 y_2 + z_1 z_2 - c^2 t_1 t_2$, which is invariant under Lorentz transformations as well. A more thorough treatment of quadrivectors can be found in Appendix A.

One of the main consequences of Lorentz transformations is a different addition law for velocities, that is expected from the invariance of the speed of light. Let us consider a particle which, as seen from reference frame O, is in (x, y, z) at time t and in $(x + \Delta x, y + \Delta y, z + \Delta z)$ at time $t + \Delta t$, thus moving with an average velocity $(V_x = \Delta x/\Delta t, V_y = \Delta y/\Delta t, V_z = \Delta z/\Delta t)$. In reference frame O' it will be instead $\Delta y' = \Delta y$, $\Delta z' = \Delta z$ and

$$\Delta x' = \frac{1}{\sqrt{1 - \frac{v^2}{c^2}}}(\Delta x - v \Delta t) , \quad \Delta t' = \frac{1}{\sqrt{1 - \frac{v^2}{c^2}}} \left(\Delta t - \frac{v}{c^2} \Delta x \right) , \qquad (1.18)$$

from which we obtain

$$V'_x \equiv \frac{\Delta x'}{\Delta t'} = \frac{\Delta x - v\Delta t}{\Delta t - \frac{v}{c^2}\Delta x} = \frac{V_x - v}{1 - \frac{vV_x}{c^2}} \quad , \quad V'_{y/z} = \sqrt{1 - \frac{v^2}{c^2}}\frac{V_{y/z}}{1 - \frac{vV_x}{c^2}} \quad (1.19)$$

instead of $V'_x = V_x - v$ and $V'_{y/z} = V_{y/z}$, as predicted by Galilean laws. It requires only some simple algebra to prove that, according to (1.19), if $|V| = c$ then also $|V'| = c$, as it should be to ensure the invariance of the speed of light: see in particular Problem (1.9) for more details on this point.

Lorentz transformations lead to some new phenomena. Let us suppose the moving observer to have a clock which is placed at rest in the origin, $x' = 0$, of its reference frame O'. The ends of any time interval $\Delta T'$ measured by that clock will correspond to two different events $(x'_1 = 0, t'_1)$ and $(x'_2 = 0, t'_2)$: they correspond to two different beats of the clock, with $t'_2 - t'_1 = \Delta T'$. The mentioned events have different coordinates in the rest frame O, which by (1.15) and setting $\gamma = 1/\sqrt{1 - v^2/c^2}$ are $(x_1 = \gamma v t'_1, \; t_1 = \gamma t'_1)$ and $(x_2 = \gamma v t'_2, \; t_2 = \gamma t'_2)$: they are therefore separated by a different time interval $\Delta T = \gamma \Delta T'$, which is in general dilated ($\gamma > 1$) with respect to the original one. This result, which can be summarized by saying that a moving clock seems to slow down, is usually known as *time dilatation*, and is experimentally confirmed by observing subatomic particles which spontaneously disintegrate with very well known mean life times: the mean life of moving particles increases with respect to that of particles at rest with the same law predicted for moving clocks. Notice that, going along the same lines, one can also demonstrate that a clock at rest in the origin of the reference frame O seems to slow down according to an observer in O': there is of course no paradox in having two different clocks, each slowing down with respect to the other, since, as long as both reference frames are inertial, the two clocks can be put together and directly compared (synchronized) only once. The same is not true in case at least one of the two frames is not inertial: a correct treatment of this case, including the well known *twin paradox*, cannot be done in the framework of Special Relativity and goes beyond the aim of the present notes.

Notice that time dilatation is also in agreement with what observed regarding the travel time of beam 1 in Michelson's interferometer, which is $2l/c$ when observed at rest and $2l/(c\sqrt{1 - v^2/c^2})$ when in motion. In order that the travel time of beam 2 be the same, we need the length l of the arm parallel to the direction of motion to appear reduced by a factor $\sqrt{1 - v^2/c^2}$, i.e. that an arm moving parallel to its length appear contracted. To confirm that, let us consider a segment of length L', at rest in reference frame O', where it is identified by the time lines $x'_1(t'_1) = 0$ and $x'_2(t'_2) = L'$ of its two ends. For an observer in O the two time lines appear as

$$x_1 = \frac{v t'_1}{\sqrt{1 - \frac{v^2}{c^2}}} \; , t_1 = \frac{t'_1}{\sqrt{1 - \frac{v^2}{c^2}}} \quad , \quad x_2 = \frac{L' + v t'_2}{\sqrt{1 - \frac{v^2}{c^2}}} \; , t_2 = \frac{t'_2 + \frac{vL'}{c^2}}{\sqrt{1 - \frac{v^2}{c^2}}} . \quad (1.20)$$

The length of the moving segment is measured in O as the distance between its two ends, located at the same time $t_2 = t_1$, i.e. $L = x_2 - x_1$ for $t'_2 = t'_1 - vL'/c^2$, so that

$$L = \frac{L'}{\sqrt{1 - \frac{v^2}{c^2}}} + \frac{v(t'_2 - t'_1)}{\sqrt{1 - \frac{v^2}{c^2}}} = L'\sqrt{1 - \frac{v^2}{c^2}} , \qquad (1.21)$$

thus confirming that, in general, any body appears contracted along the direction of its velocity (*length contraction*).

It is clear that previous formulae do not make sense when $v^2/c^2 > 1$, therefore we can conclude that it is not possible to have systems or signals moving faster than light. Last consideration leads us to a simple and useful interpretation of the length of a quadrivector. Let us take two different events in space-time, $(\boldsymbol{x}_1, ct_1)$ and $(\boldsymbol{x}_2, ct_2)$: their distance $(\Delta\boldsymbol{x}, c\Delta t)$ is also a quadrivector. If $|\Delta\boldsymbol{x}|^2 - c^2\Delta t^2 < 0$ we say that the two events have a *space-like* distance. Any signal connecting the two events would go faster than light, therefore it is not possible to establish any causal connection between them. If $|\Delta\boldsymbol{x}|^2 - c^2\Delta t^2 = 0$ we say that the two events have a *light-like* distance[1]: they are different points on the time line of a luminous signal. If $|\Delta\boldsymbol{x}|^2 - c^2\Delta t^2 < 0$ we say that the two events have a *time-like* distance: also signals slower than light can connect them, so that a causal connection is possible. Notice that these definitions are not changed by Lorentz transformations, so that the two events are equally classified by all inertial observers. For space-like distances, it is always possible to find a reference frame in which $\Delta t = 0$, or two different frames for which the sign of Δt is opposite: no absolute time ordering between the two events is possible, that being completely equivalent to saying that they cannot be put in causal connection.

Another relevant consequence of Lorentz transformations are the new laws regulating Doppler effect for electromagnetic waves propagating in vacuum space. Let us consider a monochromatic signal propagating in reference frame O with frequency ν, wavelength $\lambda = c/\nu$ and amplitude A_0, which is described by a plane wave

$$A(\boldsymbol{x}, t) = A_0 \sin(\boldsymbol{k} \cdot \boldsymbol{x} - \omega t) \qquad (1.22)$$

where $\omega = 2\pi\nu$ and $\boldsymbol{k}$ is the wave vector ($|\boldsymbol{k}| = 2\pi/\lambda = \omega/c$). Linearity implies that the signal is described by a plane wave also in the moving reference frame O', but with a new wave vector $\boldsymbol{k}'$ and a new frequency ν'. The transformation laws for these quantities are easily found by noticing that the difference of the two phases at corresponding points in the two frames must be a space-time independent constant if the fields transforms locally. That is true only if the phase $\boldsymbol{k} \cdot \boldsymbol{x} - \omega t$ is invariant under Lorentz transformations, i.e., defining $k_0 = \omega/c$, if

$$\boldsymbol{k}' \cdot \boldsymbol{x}' - k'_0 ct' = \boldsymbol{k} \cdot \boldsymbol{x} - k_0 ct . \qquad (1.23)$$

---

[1] The scalar product which is invariant under Lorentz transformations is not positive defined, so that a quadrivector can have zero length without being exactly zero.

It is easy to verify that if (1.23) has to be true for every point $(\boldsymbol{x}, ct)$ in space-time, then also the quantity $(\boldsymbol{k}, k_0)$ must transform like a quadrivector, i.e., if O' is moving with respect to O with velocity v directed along the positive x direction, we have

$$k'_x = \frac{1}{\sqrt{1 - \frac{v^2}{c^2}}} \left(k_x - \frac{v}{c} k_0 \right) \;, \quad k'_0 = \frac{1}{\sqrt{1 - \frac{v^2}{c^2}}} \left(k_0 - \frac{v}{c} k_x \right) \;, \tag{1.24}$$

and $k'_y = k_y$, $k'_z = k_z$. The first transformation law can be checked, for instance, by rewriting (1.23) with $(\boldsymbol{x}', ct') = (1, 0, 0, 0)$ and making use of (1.15); in the same way also the other laws follows. Therefore the quantity $(\boldsymbol{k}, \omega/c)$ is indeed a quadrivector, which in particular has length zero.

We can now apply (1.24) to the Doppler effect, by considering the transformation law for the frequency ν. Let us examine the particular case in which $\boldsymbol{k} = (\pm k, 0, 0)$, i.e. the wave propagates along the direction of motion of O' (longitudinal Doppler effect): after some simple algebra one finds

$$\nu' = \nu \sqrt{\frac{1 \mp v/c}{1 \pm v/c}} \;. \tag{1.25}$$

The frequency is therefore reduced (increased) if the motion of O' is parallel (opposite) to that of the signal. The result is similar to the classical Doppler effect obtained for waves propagating in a medium, but with important differences. In particular it is impossible to distinguish the motion of the source from the motion of the observer: that is evident from (1.25), since ν and ν' can be exchanged by simply reversing $v \to -v$: that is consistent with the fact that Lorentz transformations have been derived on the basis of the experimental observation that no propagating medium (ether) exists for electromagnetic waves in vacuum.

Another relevant difference is that the frequency changes even if, in frame O, the motion of O' is orthogonal to that of the propagating signal (orthogonal Doppler effect): in that case equations (1.24) imply

$$\nu' = \nu \frac{1}{\sqrt{1 - v^2/c^2}} \;. \tag{1.26}$$

We have illustrated the geometrical consequences of Lorentz transformation laws. We now want to consider the main dynamical consequences, but before doing so it will be necessary to remind some basic concepts of classical mechanics.

1.2 Relativistic Kinematics

Classical Mechanics is governed by the *minimum action* principle. A Lagrangean $\mathcal{L}(t, q_i, \dot{q}_i)$ is usually associated with a mechanical system: it has the dimensions of an energy and is a function of time, of the coordinates q_i and of their time derivatives $\dot{q}_i$. $\mathcal{L}$ is defined but for the addition of a function like $\Delta\mathcal{L}(t, q_i, \dot{q}_i) = \sum_i \dot{q}_i \, \partial F(t, q_i)/\partial q_i + \partial F(t, q_i)/\partial t$. Given a time evolution law for the coordinates $q_i(t)$ in the time interval $t_1 \leq t \leq t_2$, we define the action:

$$A = \int_{t_1}^{t_2} dt \; \mathcal{L}(t, q_i(t), \dot{q}_i(t)) . \qquad (1.27)$$

The minimum action principle states that the equations of motion are equivalent to minimizing the action in the given time interval with the constraint of having the initial and final coordinates of the system, $q_i(t_1)$ and $q_i(t_2)$, fixed. For a non relativistic particle in one dimension a possible choice for the Lagrangean is $\mathcal{L} = (1/2)m\dot{x}^2 + \text{const}$, and it is obvious that the linear motion is the one which minimizes the action among all possible time evolutions.

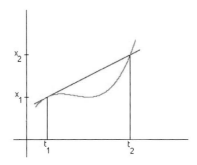

For a system of particles with positions r_i , $i = 1, ..n$, and velocities v_i, a deformation of the time evolution law: $r_i \rightarrow r_i + \delta r_i$, with $\delta r_i(t_1) = \delta r_i(t_2) = 0$, corresponds to a variation of the action :

$$\delta A = \int_{t_1}^{t_2} dt \sum_{i=1}^{n} \left[\frac{\partial \mathcal{L}}{\partial r_i} \cdot \delta r_i(t) + \frac{\partial \mathcal{L}}{\partial v_i} \cdot \delta v_i(t) \right] = \int_{t_1}^{t_2} dt \sum_{i=1}^{n} \left[\frac{\partial \mathcal{L}}{\partial r_i} - \frac{d}{dt} \frac{\partial \mathcal{L}}{\partial v_i} \right] \cdot \delta r_i(t)$$

so that the requirement that A be stationary for arbitrary variations $\delta r_i(t)$, is equivalent to the system of Lagrangean equations

$$\frac{\partial \mathcal{L}}{\partial r_i} - \frac{d}{dt} \frac{\partial \mathcal{L}}{\partial v_i} = 0 . \qquad (1.28)$$

In general it is possible to choose the Lagrangean so that the action has the same invariance properties as the equations of motion. In particular, for a free relativistic particle, the action can depend on the trajectory $q_i(t)$ in such a way that it does not change when changing the reference frame, i.e. it can be a Lorentz invariant.

Indeed, for any particular time evolution of a point-like particle, it is always possible to define an associated *proper time* as that measured by a (point-like) clock which moves (without failing) together with the particle. Since an infinitesimal proper time interval $d\tau$ corresponds to $dt = d\tau/\sqrt{1 - (v^2/c^2)}$ for an observer which sees the particle moving with velocity v, a proper time interval $\tau_2 - \tau_1$ is related to the time interval $t_2 - t_1$ measured by the observer as follows:

$$\int_{t_1}^{t_2} dt \sqrt{1 - \frac{v^2(t)}{c^2}} = \int_{\tau_1}^{\tau_2} d\tau = \tau_2 - \tau_1 \,. \tag{1.29}$$

The integral on the left hand side does not depend on the reference frame of the observer, since in any case it must correspond to the time interval $\tau_2 - \tau_1$ measured by the clock at rest with the particle: the *proper time* τ of the moving system is therefore a Lorentz invariant by definition.

The action of a free relativistic particle is a time integral depending on the evolution trajectory $\boldsymbol{r} = \boldsymbol{r}(t)$ and, in order to be invariant under Lorentz transformations, it must necessarily be proportional to the proper time

$$A = k \int_{t_1}^{t_2} dt \sqrt{1 - \frac{v^2}{c^2}} \,, \tag{1.30}$$

so that $\mathcal{L}_{free} = k \sqrt{1 - (v^2/c^2)}$. For velocities much smaller than c, we can use a Taylor expansion

$$\mathcal{L}_{free} = k \left(1 - \frac{1}{2} \frac{v^2}{c^2} - \frac{1}{8} \frac{v^4}{c^4} + \dots \right) \tag{1.31}$$

which, compared with the Lagrangean of a non relativistic free particle, gives $k = -mc^2$.

Let us now consider a scattering process involving n particles. The Lagrangean of the system at the beginning and at the end of the process, far away from when the interaction among particles is not negligible, must be equal to the sum of the Lagrangeans of the single particles, i.e.

$$\mathcal{L}(t)|_{|t| \to \infty} \to \sum_{i=1}^{n} \mathcal{L}_{free,i} = - \sum_{i=1}^{n} m_i c^2 \sqrt{1 - \frac{v_i^2}{c^2}} \,, \tag{1.32}$$

where $\mathcal{L}(t)$ is in general the complete Lagrangean describing also the interaction process and $\mathcal{L}_{free,i}$ is the Lagrangean for the i-th free particle.

If no external force are acting on the particles, $\mathcal{L}$ is invariant under translations, i.e. it does not change if the positions of all particles are translated by the same vector $\boldsymbol{a}$: $\boldsymbol{r}_i \to \boldsymbol{r}_i + \boldsymbol{a}$. This *invariance requirement* can be written as:

$$\frac{\partial \mathcal{L}}{\partial \boldsymbol{a}} = \sum_{i=1}^{n} \frac{\partial \mathcal{L}}{\partial \boldsymbol{r}_i} = 0 \,. \tag{1.33}$$

Combining this with Lagrange equations (1.28) we obtain the *conservation law*:

$$\frac{d}{dt} \sum_{i=1}^{n} \frac{\partial \mathcal{L}}{\partial \boldsymbol{v}_i} = 0 \,. \tag{1.34}$$

That means that the sum of the vector quantities $\partial \mathcal{L}/\partial \boldsymbol{v}_i$ does not change with time, hence in particular:

$$\sum_{i=1}^{n} \frac{\partial \mathcal{L}_{free,i}}{\partial v_i}\Big|_{t\to-\infty} = \sum_{i=1}^{n} \frac{\partial \mathcal{L}_{free,i}}{\partial v_i}\Big|_{t\to\infty} . \qquad (1.35)$$

In the case of relativistic particles, taking into account the following identity:

$$\frac{\partial}{\partial \boldsymbol{v}} \sqrt{1 - \frac{v^2}{c^2}} = -\frac{\boldsymbol{v}}{c^2\sqrt{1 - v^2/c^2}} \qquad (1.36)$$

and setting $\boldsymbol{v}_i|_{t\to-\infty} = \boldsymbol{v}_{i,I}$ and $\boldsymbol{v}_i|_{t\to\infty} = \boldsymbol{v}_{i,F}$, equation (1.34) reads

$$\sum_{i=1}^{n} \frac{m_i \boldsymbol{v}_{i,I}}{\sqrt{1 - v_{i,I}^2/c^2}} = \sum_{i=1}^{n} \frac{m_i \boldsymbol{v}_{i,F}}{\sqrt{1 - v_{i,F}^2/c^2}} . \qquad (1.37)$$

Invariance under translations is always related to the conservation of the total momentum of the system, hence we infer that $m\boldsymbol{v}/\sqrt{1 - v^2/c^2}$ is the generalization of *momentum* for a relativistic particle, as it is also clear by taking the limit $v/c \to 0$.

Till now we have considered the case in which the final particles coincide with the initial ones. However, in the relativistic case, particles can in general change their nature during the scattering process, melting together or splitting, losing or gaining mass, so that the final set of particles is different from the initial one. It is possible, for instance, that the collision of two particles leads to the production of new particles, or that a single particle spontaneously decays into two or more different particles. We are not interested at all, in the present context, in the specific dynamic laws regulating the interaction process; however we can say that, in absence of external forces, the invariance of the Lagrangean under spatial translation, expressed in (1.33), is valid anyway, together with the conservation law for the total momentum of the system. If we refer in particular to the initial and final states, in which the system can be described as composed by non-interacting free particles, the conservation law implies that the sum of the momenta of the initial particles be equal to the sum of the momenta of the final particles, so that (1.37) can be generalized to

$$\sum_{i=1}^{n_I} \frac{m_i^{(I)} \boldsymbol{v}_{i,I}}{\sqrt{1 - v_{i,I}^2/c^2}} = \sum_{j=1}^{n_F} \frac{m_j^{(F)} \boldsymbol{v}_{j,F}}{\sqrt{1 - v_{j,F}^2/c^2}} , \qquad (1.38)$$

where $m_i^{(I)}$ and $m_j^{(F)}$ are the masses of the n_I initial and of the n_F final particles respectively.

Similarly, if the Lagrangean does not depend explicitly on time, it is possible, using again (1.34), to derive

$$\frac{d}{dt}\mathcal{L} = \sum_i \left(\dot{\boldsymbol{v}}_i \cdot \frac{\partial \mathcal{L}}{\partial \boldsymbol{v}_i} + \boldsymbol{v}_i \cdot \frac{\partial \mathcal{L}}{\partial \boldsymbol{r}_i} \right)$$

$$= \sum_i \left(\dot{\boldsymbol{v}}_i \cdot \frac{\partial \mathcal{L}}{\partial \boldsymbol{v}_i} + \boldsymbol{v}_i \cdot \frac{d}{dt}\frac{\partial \mathcal{L}}{\partial \boldsymbol{v}_i} \right) = \frac{d}{dt} \sum_i \boldsymbol{v}_i \cdot \frac{\partial \mathcal{L}}{\partial \boldsymbol{v}_i} \qquad (1.39)$$

which is equivalent to the conservation law

$$\frac{d}{dt}\left[\sum_i \mathbf{v_i} \cdot \frac{\partial \mathcal{L}}{\partial \mathbf{v_i}} - \mathcal{L}\right] = 0 . \tag{1.40}$$

In the case of free relativistic particles the conserved quantity in (1.40) is

$$\sum_i \left(\mathbf{v_i} \cdot \frac{m_i \mathbf{v_i}}{\sqrt{1 - v_i^2/c^2}} + m_i c^2 \sqrt{1 - \frac{v_i^2}{c^2}}\right) = \sum_i \frac{m_i c^2}{\sqrt{1 - v_i^2/c^2}} . \tag{1.41}$$

In the non relativistic limit $mc^2/\sqrt{1 - v^2/c^2} \simeq mc^2 + \frac{1}{2}mv^2$ apart from terms proportional to v^4, so that (1.41) becomes the conservation law for the sum of the non relativistic kinetic energies of the particles (indeed, in the same limit, Galilean invariance laws impose conservation of mass). We can therefore consider $mc^2/\sqrt{1 - v^2/c^2}$ as the relativistic expression for the energy of a free particle, as it should be clear from the fact that invariance under time translations is always related to the conservation of the total energy. The analogous of (1.38) is

$$\sum_{i=1}^{n_I} \frac{m_i^{(I)}}{\sqrt{1 - v_{i,I}^2/c^2}} = \sum_{j=1}^{n_F} \frac{m_j^{(F)}}{\sqrt{1 - v_{j,F}^2/c^2}} . \tag{1.42}$$

It is interesting to consider how the momentum components and the energy of a particle of mass m transforms when going from one reference frame O to a new frame O' moving with velocity v_1, which is taken for instance parallel to the x axis. Let v be the modulus of the velocity of the particle and (v_x, v_y, v_z) its components in O; in order to simplify the notation, let us set $\beta = v/c$, $\beta_1 = v_1/c$, $\gamma = 1/\sqrt{1 - \beta^2}$ and $\gamma_1 = 1/\sqrt{1 - \beta_1^2}$, so that the momentum and the energy of the particle in O are

$$(p_x, p_y, p_z) = (\gamma m v_x, \gamma m v_y, \gamma m v_z) , \qquad E = \gamma m c^2 . \tag{1.43}$$

The modulus v' of the velocity of the particle in O' is easily found by using (1.19) (see Problem (1.9)):

$$\beta'^2 = \frac{v'^2}{c^2} = 1 - \frac{1}{\gamma^2 \gamma_1^2} \frac{1}{(1 - v_1 v_x/c^2)^2} , \qquad \gamma' \equiv \frac{1}{\sqrt{1 - \beta'^2}} = \gamma \gamma_1 \left(1 - \frac{v_1 v_x}{c^2}\right)$$

so that, using again (1.19),

$$p'_x = \gamma' m v'_x = \gamma_1 \gamma m \left(1 - \frac{v_1 v_x}{c^2}\right) \frac{(v_x - v_1)}{1 - v_1 v_x/c^2} = \gamma_1 \left(p_x - \frac{v_1}{c} \frac{E}{c}\right) ;$$

$$p'_{y/z} = \gamma' m v'_{y/z} = \gamma_1 \gamma \left(1 - \frac{v_1 v_x}{c^2}\right) \frac{m v_{y/z}}{\gamma_1 (1 - v_1 v_x/c^2)} = p_{y/z}; \qquad (1.44)$$

$$E' = \gamma' m c^2 = \gamma_1 \gamma m c^2 \left(1 - \frac{v_1 v_x}{c^2}\right) = \gamma_1 \left(E - \frac{v_1}{c} p_x c\right).$$

Equations (1.44) show that the quantities $(\boldsymbol{p}, E/c)$ transform in the same way as $(\boldsymbol{x}, ct)$, i.e. as the components of a quadrivector, so that the quantity $|\boldsymbol{p}|^2 - E^2/c^2$ does not change from one inertial reference frame to the other, the invariant quantity being directly linked to the mass of the particle, $|\boldsymbol{p}|^2 - E^2/c^2 = -m^2 c^2$. Moreover, given two different particles, we can construct various invariant quantities from their momenta and energies, among which $\boldsymbol{p_1} \cdot \boldsymbol{p_2} - E_1 E_2/c^2$.

If we are dealing with a system of particles, we can also build up the total momentum $\boldsymbol{P}_{tot}$ of the system, which is the sum of the single momenta, and the total energy E_{tot}, which is the sum of the single particle energies: since quadrivectors transforms linearly, the quantity $(\boldsymbol{P}_{tot}, E_{tot}/c)$, being the sum of quadrivectors, transforms as a quadrivector as well; its invariant length is linked to a quantity which is called the invariant mass (M_{inv}) of the system

$$|\boldsymbol{P}_{tot}|^2 - E_{tot}^2/c^2 \equiv -M_{inv}^2 \, c^2. \qquad (1.45)$$

It is often convenient to consider the *center-of-mass* frame for a system of particles, which is defined as the reference frame where the total momentum vanishes. It is easily verified, by using Lorentz transformations, that the center of mass moves with a relative velocity

$$\boldsymbol{v}_{cm} = \frac{\boldsymbol{P}_{tot}}{E_{tot}} c^2 \qquad (1.46)$$

with respect to an inertial frame where $\boldsymbol{P}_{tot} \neq 0$. Notice that, according to (1.45) and (1.46), the center of mass frame is well defined only if the invariant mass of the system is different from zero.

On the basis of what we have deduced about the transformation properties of energy and momentum, it is important to notice that we can identify them as the components of a quadrivector only if the energy of a particle is defined as $mc^2/\sqrt{1 - v^2/c^2}$, thus fixing the arbitrary constant usually appearing in its definition. We can then assert that a particle at rest has energy equal to mc^2. Since mass in general is not conserved[2], it may happen that part of the rest energy of a decaying particle is transformed into the kinetic energy of the final particles, or that part of the kinetic energy of two or more particles involved in a diffusion process is transformed in the rest energy of the final particles. As an example, the energy which comes out of a nuclear fission process derives

---

[2] We mean by that the sum of the masses of the single particles. The invariant mass of a system of particles defined in (1.45) is instead conserved, as a consequence of the conservation of total momentum and energy.

from the excess in mass of the initial nucleus with respect to the masses of the products of fission. Everybody knows the relevance that this simple remark has acquired in the recent past.

A particular discussion is required for particles of vanishing mass, like the photon. While, according to (1.37) and (1.41), such particles seem to have vanishing momentum and energy, a more careful look shows that the limit $m \to 0$ can be taken at constant momentum p, if at the same time the speed of the particle tends to c according to $v = c \, (1 + m^2 p^2 / c^2)^{-1/2}$. It can be easily verified that in the same limit $E \to pc$, in perfect agreement with $p^2 - E^2/c^2 = -m^2 c^2$.

Our considerations on conservation and transformation properties of energy and momentum permit to fix in a quite simple way the kinematic constraints related to a diffusion process: let us illustrate this point with an example.

In relativistic diffusion processes it is possible to produce new particles starting from particles which are commonly found in Nature. The collision of two hydrogen nuclei (protons), which have a mass $m \simeq 1.67 \ 10^{-27}$ Kg, can generate the particle π, which has a mass $\mu \simeq 2.4 \ 10^{-28}$ Kg. Technically, some protons are accelerated in the reference frame of the *laboratory* till one obtains a beam of momentum P, which is then directed against hydrogen at rest. That leads to proton–proton collisions from which, apart from the already existing protons, also the π particles emerge (it is possible to describe schematically the reaction as $p + p \to p + p + \pi$). A natural question regards the minimum momentum or energy of the beam particles needed to produce the reaction: in order to get an answer it is convenient to consider this problem as seen from the reference frame of the center of mass, in which the two particles have opposite momenta, which we suppose to be parallel, or anti-parallel, to the x axis: $P_1 = -P_2$, and equal energies $E_1 = \sqrt{P_1^2 c^2 + m^2 c^4} = E_2$. In this reference frame the total momentum vanishes and the total energy is $E = 2E_1$. Conservation of momentum and energy constrains the sum of the three final particle momenta to vanish, and the sum of their energy to be equal to E. The required energy is minimal if all final particles are produced at rest, the kinematical constraint on the total momentum being automatically satisfied in that case (this is the advantage of doing computations in the center of mass frame). We can then conclude that the minimum value of E in the center of mass is $E_{min} = (2m + \mu)c^2$. However that is not exactly the answer to our question: we have to find the value of the energy of the beam protons corresponding to a total energy in the center of mass equal to E_{min}. That can be done by noticing that in the center of mass both colliding protons have energy $E_{min}/2$, so that we can compute the relative velocity βc between the center of mass and the laboratory as that corresponding to a Lorentz transformation leading from a proton at rest to a proton with energy $E_{min}/2$, i.e. solving the equation

$$\frac{1}{\sqrt{1-\beta^2}} = \frac{E_{min}}{2mc^2} = \frac{2m+\mu}{2m} \ .$$

While, as said above, the total momentum of the system vanishes in the center of mass frame and the total energy equals E_{min}, in the laboratory the total momentum is obtained by Lorentz transformation as

$$P_L = \frac{\beta}{\sqrt{1-\beta^2}}\frac{E_{min}}{c} = \sqrt{\frac{1}{1-\beta^2}-1}\,\frac{E_{min}}{c} = (2m+\mu)c\sqrt{\frac{(2m+\mu)^2}{4m^2}-1}$$

$$= \frac{2m+\mu}{2m}c\sqrt{4m\mu+\mu^2} \ .$$

This is also the answer to our question, since in the laboratory all momentum is carried by the proton of the beam.

An alternative way to obtain the same result, without making explicit use of Lorentz transformations, is to notice that, if E_L is the total energy in the laboratory, $P_L^2 - E_L^2/c^2$ is an invariant quantity, which is therefore equal to the same quantity computed in the center of mass, so that

$$P_L^2 - \frac{E_L^2}{c^2} = -(2m+\mu)^2 c^2 \ .$$

Writing E_L as the sum of the energy of the proton in the beam, which is $\sqrt{P_L^2 c^2 + m^2 c^4}$, and of that of the proton at rest, which is mc^2, we have the following equation for P_L:

$$P_L^2 - \frac{1}{c^2}\left[\sqrt{P_L^2 c^2 + m^2 c^4} + mc^2\right]^2 = -(2m+\mu)^2 c^2 \ .$$

which finally leads to the same result obtained above.

Suggestions for Supplementary Readings

- C. Kittel, W. D. Knight, M. A. Ruderman: *Mechanics - Berkeley Physics Course*, volume 1 (Mcgraw-Hill Book Company, New York 1965)
- L. D. Landau, E. M. Lifchitz: *The Classical Theory of Fields - Course of theoretical physics*, volume 2 (Pergamon Press, London 1959)
- J. D. Jackson: *Classical Electrodynamics*, 3d edn (John Wiley, New York 1998)
- W. K. H. Panowsky, M. Phillips: *Classical Electricity and Magnetism*, 2nd edn (Addison-Wesley Publishing Company Inc., Reading 2005)

Problems

1.1. A spaceship of length $L_0 = 150$ m is moving with respect to a space station with a speed $v = 2\ 10^8$ m/s. What is the length L of the spaceship as measured by the space station?

Answer: $L = L_0\sqrt{1 - v^2/c^2} \simeq 112$ m .

1.2. How many years does it take for an atomic clock (with a precision of one part over 10^{15}), which is placed at rest on Earth, to lose one second with respect to an identical clock placed on the Sun? (Hint: apply Lorentz transformation laws as if both reference frames were inertial, with a relative velocity $v \simeq 3\ 10^4$ m/s $\simeq 10^{-4}\ c$).

Answer: $T = \left(1 - \sqrt{1 - v^2/c^2}\right)^{-1}$ s $\simeq 2c^2/v^2$ s $\simeq 6.34$ years .

1.3. An observer casts a laser pulse with a frequency $\nu = 10^{15}$ Hz against a mirror, which is moving with a speed $v = 5\ 10^7$ m/s opposite to the direction of the pulse, and whose surface is orthogonal to it. The observer then measures the frequency ν' of the pulse coming back after being reflected by the mirror. What is the value of ν'?

Answer: Since reflection leaves the frequency unchanged only in the rest frame of the mirror, two different longitudinal Doppler effects have to be taken into account, that of the original pulse with respect to the mirror and that of the reflected pulse with respect to the observer, therefore $\nu' = \nu\ (1 + v/c)/(1 - v/c) = 1.4\ 10^{15}$ Hz .

1.4. Fizeau's Experiment

In the experiment described in the figure, a light beam of frequency $\nu = 10^{15}$ Hz, produced by the source S, is split into two distinct beams which go along two different paths belonging to a rectangle of sides $L_1 = 10$ m and $L_2 = 5$ m. They recombine, producing interference in the observation point O, as illustrated in the figure. The rectangular path is contained in a tube T filled with a liquid having refraction index $n = 2$, so that the speed of light in that liquid is $v_c \simeq 1.5\ 10^8$ m/s. If the liquid is moving counterclockwise around the tube with a velocity 0.3 m/s, the speed of the light beams along the two different paths changes, together with their wavelength, which is constrained by the equation $v_c = \lambda\nu$ (the frequency ν instead does not change and is equal to that of the original beam). For that reason the two beams recombine in O with a phase difference $\Delta\phi$, which is different from zero (the phase accumulated by each beam is given by 2π times the number of wavelengths contained in the total path). What is the value of $\Delta\phi$? Compare the result with what would have been obtained using Galilean transformation laws.

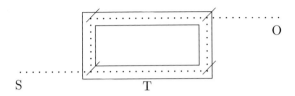

Answer: Calling $L = L_1 + L_2 = 15$ m the total path length inside the tube for each beam, and using Einstein's laws for adding velocities, one finds

$$\Delta\phi = 4\pi\nu L v \frac{n^2 - 1}{c^2 - n^2 v^2} \simeq 4\pi\nu L v (n^2 - 1)/c^2 \simeq 1.89 \text{ rad} .$$

Instead, Galileo's laws would lead to

$$\Delta\phi = 4\pi\nu L v \frac{n^2}{c^2 - n^2 v^2} ,$$

a result which does not make sense since it is different from zero also when the tube is empty: in that case, indeed, Galilean law would imply that the tube is still filled with a "rotating ether".

1.5. My clock gains one second every year with respect to the official time. Trying to correct for that, I put the clock in uniform rotation at one end of a strong rope of length $L = 2$ m, with an angular velocity ω. What value should I choose for ω to get the correct time?

Answer: Time intervals of my clock are related to official time intervals by $\Delta t_{cl} = (1 - \epsilon)\Delta t_{of}$ with $\epsilon = (1 \text{ s})/(1 \text{ year}) \simeq 3.1 \ 10^{-8}$. We require $\Delta t'_{cl} = \Delta t_{of}$ where $\Delta t'_{cl} = \gamma \Delta t_{cl}$ is the time interval as measured by the rotating clock, with $\gamma = 1/\sqrt{1 - \omega^2 L^2/c^2}$ (we have used relativistic time dilatation and have been a bit cavalier since a rotating clock is not exactly an inertial reference frame). Since $\epsilon \ll 1$, we obtain $\omega^2 L^2 \simeq 2\epsilon c^2$, hence $\omega = v/L \simeq 3.7 \ 10^4$ rad/s.

1.6. A flux of particles, each carrying an electric charge $q = 1.6 \ 10^{-19}$ C, is moving along the x axis with a constant velocity $v = 0.9 \ c$. If the total carried current is $I = 10^{-9}$ A, what is the linear density of particles, as measured in the reference frame at rest with them?

Answer: If d_0 is the distance among particles in their rest frame, the distance measured in the laboratory appears contracted and equal to $d = \sqrt{1 - v^2/c^2} d_0$. The electric current is given by $I = d^{-1} v q$, hence the particle density in the rest frame is

$$d_0^{-1} = \sqrt{1 - v^2/c^2} \frac{I}{vq} \simeq 10.1 \text{ particles/m} .$$

1.7. The point of an arrow is an isosceles triangle having its height h twice the length l of its base, which is orthogonal to the arrow. With what velocity should the arrow be thrown so that its point appears to be an equilateral triangle?

Answer: Assuming the arrow to be parallel to its velocity, the height of the triangle appears to be contracted $h' = \sqrt{1 - v^2/c^2}\, h$, while its base stays unchanged $l' = l = h/2$. The request $h' = \sqrt{3}l'/2$ (equilater triangle) is then satisfied if $v^2/c^2 = 13/16$, i.e. $v \simeq 2.7\ 10^8$ m/s.

1.8. A particle is moving with a velocity $c/2$ along the positive direction of the line $y = x$. What are the components of the velocity of the particle for an observer moving with a speed $V = 0.99\,c$ along the x axis?

Answer: We have $v_x = v_y = c/(2\sqrt{2})$ in the original system. After applying relativistic laws for the addition of velocities we find

$$v'_x = \frac{v_x - V}{1 - v_x V/c^2} \simeq -0.979c\ ; \quad v'_y = \sqrt{1 - V^2/c^2}\,\frac{v_y}{1 - v_x V/c^2} \simeq 0.0767c\,.$$

1.9. A particle is moving with a speed of modulus v and components (v_x, v_y, v_z). What is the modulus v' of the velocity for an observer moving with a speed w along the x axis? Comment the result as v and/or w approach c.

Answer: Applying the relativistic laws for the addition of velocities we find

$$v'^2 = v'^2_x + v'^2_y + v'^2_z = \frac{(v_x - w)^2 + (1 - w^2/c^2)(v^2_z + v^2_y)}{(1 - v_x w/c^2)^2}$$

and after some simple algebra

$$v'^2 = c^2 \left(1 - \frac{(1 - v^2/c^2)(1 - w^2/c^2)}{(1 - v_x w/c^2)^2} \right).$$

It is interesting to notice that as v and/or w approach c, also v' approaches c (from below): that verifies that a body moving with $v = c$ moves with the same velocity in every reference frame (invariance of the speed of light).

1.10. Two spaceships, moving along the same course with the same velocity $v = 0.98\,c$, pass space station Alpha, which is placed on their course, at the same hour of two successive days. On each of the two spaceships a radar permits to know the distance from the other spaceship: what value does it measure?

Answer: In the reference frame of space station Alpha, the two spaceships stay at the two ends of a segment of length $L = vT$ where $T = 1$ day. That distance is reduced by a factor $\sqrt{1 - v^2/c^2}$ with respect to the distance L_0 among the spaceships as measured in their rest frame. We have therefore

$$L_0 = \frac{1}{\sqrt{1 - v^2/c^2}} vT \simeq 1.28\ 10^{14}\ \text{m}.$$

1.11. We are on the course of a spaceship moving with a constant speed v while emitting electromagnetic pulses which, in the rest frame of the spaceship, are equally spaced in time. We receive a pulse every second while the spaceship is approaching us, and a pulse every two seconds while the spaceship is leaving us. What is the speed of the spaceship?

Answer: It can be easily checked that the frequency of pulses changes according to the longitudinal Doppler effect, so that

$$\frac{1 + v/c}{1 - v/c} = 2,$$

hence $v = 1/3\ c$.

1.12. During a Star Wars episode, space station Alpha detects an enemy spaceship approaching it from a distance $d = 10^8$ Km at a speed $v = 0.9\ c$, and at the same time the station launches a missile of speed $v' = 0.95\ c$ to destroy it. As soon as the enemy spaceship detects the electromagnetic pulses emitted by the missile, it launches against space station Alpha the same kind of missile, therefore moving at a speed $0.95\ c$ in the rest frame of the spaceship.
How much time do the inhabitants of space station Alpha have, after having launched their missile, to leave the station before it is destroyed by the second missile?

Answer: Let us make computations in the reference frame of space station Alpha. Setting to zero the launching time of the first missile, the enemy spaceship detects it and launches the second missile at time $t_1 = d/(v+c)$ and when it is at a distance $x_1 = cd/(v+c)$ from the space station. The second missile approaches Alpha with a velocity $V = (v + v')/(1 + vv'/c^2) \simeq .9973\ c$, therefore it hits the space station at time $t_2 = t_1 + x_1/V \simeq 352$ s.

1.13. A particle moves in one dimension with an acceleration which is constant and equal to a in the reference frame instantaneously at rest with it. Determine, for $t > 0$, the time line $x(t)$ of the particle in the reference frame of the laboratory, where it is placed at rest in $x = 0$ at time $t = 0$.

Answer: Let us call τ the proper time of the particle, synchronized so that $\tau = 0$ when $t = 0$, The relation between proper and laboratory time is given by

$$d\tau = \sqrt{1 - v^2(t)/c^2}\ dt$$

where $v(t) = dx/dt$ is the particle velocity in the laboratory. Let us also introduce the *quadrivelocity* $(d\boldsymbol{x}/d\tau, d(ct)/d\tau)$, which transforms as a quadrivector since $d\tau$ is a Lorentz invariant. We are interested in particular in its spatial component in one dimension, $u \equiv dx/d\tau$, which is related to v by the following relations

$$u = \frac{v}{\sqrt{1 - v^2/c^2}}, \qquad v = \frac{u}{\sqrt{1 + u^2/c^2}}, \qquad \sqrt{1 + \frac{u^2}{c^2}}\sqrt{1 - \frac{v^2}{c^2}} = 1.$$

To derive the equation of motion, let us notice that, in the frame instantaneously at rest with the particle, the velocity goes from 0 to $ad\tau$ in the interval $d\tau$, so that v changes into $(v + ad\tau)/(1 + vad\tau/c^2)$ in the same interval, meaning that $dv/d\tau = a(1 - v^2/c^2)$. Using previous equations, it is easy to derive

$$\frac{du}{dt} = \frac{du}{dv}\frac{dv}{d\tau}\frac{d\tau}{dt} = a$$

which can immediately be integrated, with the initial condition $u(0) = v(0) = 0$, as $u(t) = at$. Using the relation between u and v we have

$$v(t) = \frac{at}{\sqrt{1 + a^2t^2/c^2}} .$$

That gives the variation of velocity, as observed in the laboratory, for a uniformly accelerated motion: for $t \ll c/a$ one recovers the non-relativistic result, while for $t \gg c/a$ the velocity reaches asymptotically that of light. The dependence of v on t can be finally integrated, using the initial condition $x(0) = 0$, giving

$$x(t) = \frac{c^2}{a}\left(\sqrt{1 + \frac{a^2t^2}{c^2}} - 1\right) .$$

1.14. Spaceship A is moving with respect to space station S with a velocity $2.7\ 10^8$ m/s. Both A and S are placed in the origin of their respective reference frames, which are oriented so that the relative velocity of A is directed along the positive direction of both x axes; A and S meet at time $t_A = t_S = 0$. Space station S detects an event, corresponding to the emission of luminous pulse, in $x_S = 3\ 10^{13}$ m at time $t_S = 0$. An analogous but distinct event is detected by spaceship A, with coordinates $x_A = 1.3\ 10^{14}$ m , $t_A = 2.3\ 10^3$ s. Is it possible that the two events have been produced by the same moving body?

Answer: The two events may have been produced by the same moving body only if they have a time-like distance, otherwise the unknown body would go faster than light. After obtaining the coordinates of the two events in the same reference frame, one obtains $\Delta x = 6.11\ 10^{13}$ m and $c\Delta t = 6.28\ 10^{13}$ m $> \Delta x$: the two events may indeed have been produced by the same body moving at a speed $\Delta x/\Delta t \simeq 0.97\ c$.

1.15. We are moving towards a mirror with a velocity v orthogonal to its surface. We send an electromagnetic pulse of frequency $\nu = 10^9$ Hz towards the mirror, along the same direction of our motion. After 2 seconds we receive a reflected pulse of frequency $\nu' = 1.32\ 10^9$ Hz. How many seconds are left before our impact with the mirror?

Answer: We can deduce our velocity with respect to the mirror by using the double longitudinal Doppler effect: $v = \beta c$ with $(1 + \beta)/(1 - \beta) = 1.32$, hence $v = 0.138\ c$. One second after we emit our pulse, the mirror is placed at one light-second from us, therefore the impact will take place after $\Delta T = 1/0.138$ s $= 7.25$ s, that means 6.25 s after we receive the reflected pulse.

1.16. A particle μ of mass $m = 1.7 \ 10^{-28}$ Kg and carrying the same electric charge as the electron has a mean life time $\tau = 10^{-6}$ s when it is at rest. The particle is accelerated instantaneously through a potential gap $\Delta V = 10^8$ V. What is the expected life time of the particle, in the laboratory, after the acceleration?

Answer: $t = \tau \ (mc^2 + e\Delta V)/mc^2 \simeq 1.96 \ 10^{-6}$ s.

1.17. The energy of a particle is equal to $2.5 \ 10^{-12}$ J, its momentum is $7.9 \ 10^{-21}$ N s. What are its mass m and velocity v?

Answer: $m = \sqrt{E^2 - c^2 p^2}/c^2 \simeq 8.9 \ 10^{-30}$ Kg , $v = pc^2/E \simeq 2.84 \ 10^8$ m/s.

1.18. A spaceship with an initial mass $M = 10^3$ Kg is boosted by a photonic engine: a light beam is emitted opposite to the direction of motion, with a power $W = 10^{15}$ W, as measured in the spaceship rest frame. What is the derivative of the spaceship rest mass with respect to its proper time? What is the spaceship acceleration a in the frame instantaneously at rest with it?

Answer: The engine power must be subtracted from the spaceship energy, which is Mc^2 in its rest frame, hence $dM/dt = -W/c^2 \simeq 1.1 \ 10^{-2}$ Kg/s. Since the particles emitted by the engine are photons, they carry a momentum equal to $1/c$ times their energy, hence

$$a(\tau) = \frac{W}{c(M - W\tau/c^2)} \ .$$

1.19. A spaceship with an initial mass $M = 3 \ 10^4$ Kg is boosted by a photonic engine of constant power, as measured in the spaceship rest frame, equal to $W = 10^{13}$ W. If the spaceship moves along the positive x direction and leaves the space station at $\tau = 0$, compute its velocity with respect to the station reference frame (which is assumed to be inertial) as a function of the spaceship proper time.

Answer: According to the solution of previous problem (1.18), the spaceship acceleration a in the frame instantaneously at rest with it is

$$a(\tau) = \frac{W}{cM(\tau)} = \frac{W}{c(M - \tau W/c^2)} = \frac{a_0}{1 - \alpha\tau} \ , \qquad a_0 \simeq 1.1 \ \text{m/s}^2 \ , \qquad \alpha \simeq 3.7 \ 10^{-9} \ \text{s}^{-1}$$

Recalling from the solution of problem (1.13) that

$$v = \frac{u}{\sqrt{1 + \frac{u^2}{c^2}}} \ , \qquad \frac{du}{d\tau} = a(\tau)\sqrt{1 + \frac{u^2}{c^2}} \ ,$$

where u is the x-component of the quadrivelocity, we can easily integrate last equation

$$\frac{du}{\sqrt{1 + \frac{u^2}{c^2}}} = \frac{a_0 d\tau}{1 - \alpha\tau}$$

obtaining

$$\frac{u}{c} = \sinh\left(-\frac{a_0}{\alpha c}\ln(1 - \alpha\tau)\right).$$

Expressing v/c as a function of u/c we finally get

$$\frac{v}{c} = \tanh\left(-\frac{a_0}{\alpha c}\ln(1 - \alpha\tau)\right) = \frac{1 - (1 - \alpha\tau)^{2a_0/\alpha c}}{1 + (1 - \alpha\tau)^{2a_0/\alpha c}}.$$

1.20. An electron–proton collision can give rise to a fusion process in which all available energy is transferred to a neutron. As a matter of fact, there is a neutrino emitted whose energy and momentum in the present situation can be neglected. The proton rest energy is $0.938 \ 10^9$ eV, while those of the neutron and of the electron are respectively $0.940 \ 10^9$ eV and $5 \ 10^5$ eV. What is the velocity of the electron which knocking into a proton at rest may give rise to the process described above?

Answer: Notice that we are not looking for a minimum electron energy: since, neglecting the final neutrino, the final state is a single particle state, its invariant mass is fixed and equal to the neutron mass. That must be equal to the invariant mass of the initial system of two particles, leaving no degrees of freedom on the possible values of the electron energy: only for one particular value v_e of the electron velocity the reaction can take place.

A rough estimate of v_e can be obtained by considering that the electron energy must be equal to the rest energy difference $(m_n - m_p)\,c^2 = (0.940 - 0.938)\ 10^9$ eV plus the kinetic energy of the final neutron. Therefore the electron is surely relativistic and $(m_n - m_p)c$ is a reasonable estimate of its momentum: it coincides with the neutron momentum which is instead non-relativistic $((m_n - m_p)c \ll m_n c)$. The kinetic energy of the neutron is thus roughly $(m_n - m_p)^2 c^2/(2m_n)$, hence negligible with respect to $(0.940 - 0.938)\ 10^9$ eV. The total electron energy is therefore, within a good approximation, $E_e \simeq 2\ 10^6$ eV, and its velocity is $v_e = c\sqrt{1 - m_e^2/E_e^2} \simeq 2.9 \ 10^8$ m/s. The exact result is obtained by writing $E_e = (m_n^2 - m_p^2 - m_e^2)c^2/(2m_p)$, which differs by less than 0.1 % from the approximate result.

1.21. A system made up of an electron and a positron, which is an exact copy of the electron but with opposite charge (i.e. its antiparticle), annihilates, while both particles are at rest, into two photons. The mass of the electron is $m_e \simeq 0.9 \ 10^{-31}$ Kg: what is the wavelength of each outgoing photon? Explain why the same system cannot decay into a single photon.

Answer: The two photons carry momenta of modulus mc which are opposite to each other in order to conserve total momentum : their common wavelength is therefore, as we shall see in next Chapter, $\lambda = h/(mc) \simeq 4.2 \ 10^{-13}$ m . A single photon should carry zero momentum since the initial system is at rest, but then energy could not be conserved; more in general the initial invariant mass of the system, which is $2m_e$, cannot fit the invariant mass of a single photon, which is always zero.

1.22. A piece of copper of mass $M = 1$ g, is heated from $0°C$ up to $100°C$. What is the mass variation ΔM if the copper specific heat is $C_{Cu} = 0.4$ J/g$°C$?

Answer: The piece of copper is actually a system of interacting particles whose mass is defined as the invariant mass of the system. That is proportional to the total energy if the system is at rest, see equation (1.45). Therefore $\Delta M = C_{\mathrm{Cu}} \, \Delta T/c^2 \simeq$ $4.45 \; 10^{-16}$ Kg .

1.23. A photon of energy E knocks into an electron at rest producing a final state composed of an electron–positron pair plus the initial electron: all three final particles have the same momentum. What is the value of E and the common momentum p of the final particles?

Answer: $E = 4mc^2 \simeq 3.3 \; 10^{-13}$ J , $p = E/3c = 4/3 \; mc \simeq 3.6 \; 10^{-22}$ N/m .

1.24. A particle of mass $M = 10^{-27}$ Kg decays, while at rest, into a particle of mass $m = 4 \; 10^{-28}$ Kg plus a photon. What is the energy E of the photon?

Answer: The two outgoing particles must have opposite momenta with an equal modulus p to conserve total momentum. Energy conservation is then written as $Mc^2 = \sqrt{m^2 c^4 + p^2 c^2} + pc$, so that

$$E = pc = \frac{M^2 - m^2}{2M} c^2 = 0.42 \; Mc^2 \simeq 3.78 \; 10^{-11} \; \mathrm{J} \simeq 2.36 \; 10^8 \; \mathrm{MeV} \ .$$

1.25. A particle of mass $M = 1 \, \mathrm{GeV}/c^2$ and energy $E = 10 \, \mathrm{GeV}$ decays into two particles of equal mass $m = 490 \, \mathrm{MeV}$. What is the maximum angle that each of the two outgoing particles may form, in the laboratory, with the trajectory of the initial particle?

Answer: Let $\hat{x}$ be the direction of motion of the initial particle, and x–y the decay plane: this is defined as the plane containing both the initial particle momentum and the two final momenta, which are indeed constrained to lie in the same plane by total momentum conservation. Let us consider one of the two outgoing particles: in the center of mass frame it has energy $\epsilon = Mc^2/2 = 0.5 \, \mathrm{GeV}$ and a momentum $p_x = p \cos \theta$, $p_y = p \sin \theta$, with θ being the decay angle in the center of mass frame and

$$p = c \sqrt{\frac{M^2}{4} - m^2} \simeq \frac{0.1 \, \mathrm{GeV}}{c} \ .$$

The momentum components in the laboratory are obtained by Lorentz transformations with parameters $\gamma = (\sqrt{1 - v^2/c^2})^{-1} = E/(Mc^2) = 10$ and $\beta = v/c = \sqrt{1 - 1/\gamma^2} \simeq 0.995$,

$$p'_y = p_y = p \sin \theta$$
$$p'_x = \gamma(p \cos \theta + \beta \epsilon) \ .$$

It can be easily verified that, while in the center of mass frame the possible momentum components lie on a circle of radius p centered in the origin, in the laboratory p'_x and p'_y lie on an ellipse of axes γp and p, centered in $(\gamma \beta \epsilon, 0)$. If θ' is the angle formed in the laboratory with respect to the initial particle trajectory and defining $\alpha \equiv \beta \epsilon/p \simeq 5$, we can write

$$\tan \theta' = \frac{p'_y}{p'_x} = \frac{1}{\gamma} \frac{\sin \theta}{\cos \theta + \alpha}.$$

If $\alpha > 1$ the denominator is always positive, so that $\tan \theta'$ is limited and $|\theta'| < \pi/2$, i.e. the particle is always forward emitted, in the laboratory, with a maximum possible angle which can be computed by solving $d\tan \theta'/d\theta = 0$; that can also be appreciated pictorially by noticing that if $\alpha > 1$ the ellipse containing the possible momentum components does not contain the origin. Finally one finds

$$\cos \theta = -1/\alpha \; ; \quad \theta'_{max} = \tan^{-1} \left(\frac{1}{\gamma} \frac{1}{\sqrt{\alpha^2 - 1}} \right) \simeq 0.02 \text{ rad} .$$

1.26. A particle of mass $M = 10^{-27}$ Kg, which is moving in the laboratory with a speed $v = 0.99\,c$, decays into two particles of equal mass $m = 3\ 10^{-28}$ Kg. What is the possible range of energies (in GeV) which can be detected in the laboratory for each of the outgoing particles?

Answer: In the center of mass frame the outgoing particles have equal energy and modulus of the momentum completely fixed by the kinematic constraints: $E = Mc^2/2$ and $P = c\sqrt{M^2/4 - m^2}$. The only free variable is the decaying angle θ, measured with respect to the initial particle trajectory, which however results in a variable energy E' in the laboratory system. Indeed by Lorentz transformations $E' = \gamma(E + v \cos \theta P)$, with $\gamma = (1 - v^2/c^2)^{-1/2}$, hence

$$E'_{\text{max/min}} = \gamma \left(\frac{Mc^2}{2} \pm \frac{v}{c} Pc \right) \; ; \quad E'_{\text{max}} \simeq 3.567 \text{ GeV} , \; E'_{\text{min}} \simeq 0.41 \text{ GeV}$$

1.27. A particle of rest energy $Mc^2 = 10^9$ eV, which is moving in the laboratory with momentum $p = 5\ 10^{-18}$ N s, decays into two particles of equal mass $m = 2\ 10^{-28}$ Kg. In the center of mass frame the decay direction is orthogonal to the trajectory of the initial particle. What is the angle between the trajectories of the outgoing particles in the laboratory?

Answer: Let $\hat{x}$ be the direction of the initial particle and $\hat{y}$ the decay direction in the center of mass frame, $\hat{y} \perp \hat{x}$. For the process described in the text, $\hat{x}$ is a symmetry axis, hence the outgoing particles will form the same angle θ also in the laboratory. For one of the two particles we can write $p_x = p/2$ by momentum conservation in the laboratory, and $p_y = c\sqrt{M^2/4 - m^2}$ by energy conservation. Finally, the angle between the two particles is $2\theta = 2 \operatorname{atan}(p_y/p_x) \simeq 0.207$ rad .

1.28. Compton Effect
A photon of wavelength λ knocks into an electron at rest. After the elastic collision, the photon moves in a direction forming an angle θ with respect to its original trajectory. What is the change $\Delta\lambda \equiv \lambda' - \lambda$ of its wavelength as a function of θ?

Answer: Let $\mathbf{q}$ and $\mathbf{q}'$ be respectively the initial and final momentum of the photon, and $\mathbf{p}$ the final momentum of the electron. As we shall discuss in next Chapter, the photon momentum is related to its wavelength by the relation $q \equiv |\mathbf{q}| = h/\lambda$, where

h is Planck's constant. Total momentum conservation implies that q, q' and p must lie in the same plane, which we choose to be the $x - y$ plane, with the x-axis parallel to the photon initial trajectory. Momentum and energy conservation lead to:

$$p_x = q - q' \cos \theta \,,$$

$$p_y = q' \sin \theta \,,$$

$$qc + m_e c^2 - q'c = \sqrt{m_e c^4 + p_x^2 c^2 + p_y^2 c^2} \,.$$

Substituting the first two equations into the third and squaring both sides of the last we easily arrive, after some trivial simplifications, to $m_e(q - q') = qq'(1 - \cos \theta)$, which can be given in terms of wavelengths as follows

$$\Delta \lambda \equiv \lambda' - \lambda = \frac{h}{m_e c}(1 - \cos \theta) \,.$$

The difference is always positive, since part of the photon energy, depending on the diffusion angle θ, is always transferred to the electron. This phenomenon, known as *Compton effect*, is not predicted by the classical theory of electromagnetic waves and is an experimental proof of the corpuscular nature of radiation. Notice that, while the angular distribution of outgoing photons can only be predicted on the basis of the quantum relativistic theory, i.e. Quantum Electrodynamics, the dependence of $\Delta \lambda$ on θ that we have found is only based on relativistic kinematics and can be used to get an experimental determination of h. The coefficient $h/(m_e c)$ is known as *Compton wavelength*, which for the electron is of the order of 10^{-12} m, so that the effect is not detectable ($\Delta \lambda / \lambda \simeq 0$) for visible light.

1.29. A spinning-top, which can be described as a rigid disk of mass $M = 10^{-1}$ Kg, radius $R = 5 \, 10^{-2}$ m and uniform density, starts rotating with angular velocity $\Omega = 10^3$ rad/s . What is the energy variation of the spinning-top due to rotation, as seen from a reference frame moving with a relative speed $v = 0.9 \, c$ with respect to it?

Answer: The speed of the particles composing the spinning-top is surely non-relativistic in their frame since it is limited by $\Omega R = 50$ m/s $\simeq 1.67 \, 10^{-7}$ c. In that system the total energy is therefore, apart from corrections of order $(\Omega R/c)^2$, $E_{tot} = Mc^2 + I\Omega^2/2$, where I is the moment of inertia, $I = MR^2/2$. Lorentz transformations yield in the moving system

$$E'_{tot} = \frac{1}{\sqrt{1 - v^2/c^2}} E_{tot} = \frac{1}{\sqrt{1 - v^2/c^2}} \left(Mc^2 + \frac{1}{2} I\Omega^2 \right) \,,$$

to be compared with energy observed in absence of rotation, $Mc^2/\sqrt{1 - v^2/c^2}$. Therefore, the energy variation due to rotation in the moving frame is $\Delta E' = I\Omega^2/(2\sqrt{1 - v^2/c^2}) \simeq 143$ J .

1.30. A photon with energy $E_1 = 10^4$ eV is moving along the positive x direction; a second photon with energy $E_2 = 2 \, E_1$ is moving along the positive y direction. What is the velocity $\mathbf{v}_{cm}$ of their center of mass frame?

Answer: Recalling equation (1.46), it is easily found that $v_{cm}^x = 1/3 \, c$, $v_{cm}^y = 2/3 \, c$ and $|\mathbf{v}_{cm}| = \sqrt{5}/3 \, c$.

1.31. A particle of mass M decays, while at rest, into three particles of equal mass m. What is the maximum and minimum possible energy for each of the outgoing particles?

Answer: Let E_1, E_2, E_3 and $\boldsymbol{p}_1$, $\boldsymbol{p}_2$, $\boldsymbol{p}_3$ be respectively the energies and the momenta of the three outgoing particles. We have to find, for instance, the maximum and minimum value of E_1 which are compatible with the constraints $\boldsymbol{p}_1 + \boldsymbol{p}_2 + \boldsymbol{p}_3 = 0$ and $E_1 + E_2 + E_3 = M\ c^2$. The minimum value is realized when the particle is produced at rest, $E_{1min} = m\ c^2$, implying that the other two particles move with equal and opposite momenta. Finding the maximum value requires some more algebra. From $E_1^2 = m^2\ c^4 + p_1^2\ c^2$ and momentum conservation we obtain

$$E_1^2 = m^2\ c^4 + |\boldsymbol{p}_2 + \boldsymbol{p}_3|^2\ c^2 = m^2\ c^4 + (E_2 + E_3)^2 - \mu^2\ c^4$$

where μ is the invariant mass of particles 2 and 3, $\mu^2\ c^4 = (E_2 + E_3)^2 - |\boldsymbol{p}_2 + \boldsymbol{p}_3|^2\ c^2$. Applying energy conservation, $E_2 + E_3 = (M\ c^2 - E_1)$, last equation leads to

$$E_1 = \frac{1}{2M\ c^2}\left(m^2\ c^4 + M^2\ c^4 - \mu^2\ c^4\right).$$

We have written E_1 as a function of μ^2: E_{1max} corresponds to the minimum possible value for the invariant mass of the two remaining particles. On the other hand it can be easily checked that, for a system made up of two or more massive particles, the minimum possible value of the invariant mass is equal to the sum of the masses and is attained when the particles are at rest in their center of mass frame, meaning that the particles move with equal velocities in any other reference frame. Therefore $\mu_{min} = 2m$ and $E_{1max} = (M^2 - 3m^2)\ c^2/(2M)$: this value is obtained in particular for $\boldsymbol{p}_2 = \boldsymbol{p}_3$.

1.32. A proton beam is directed against a laser beam coming from the opposite direction and having wavelength $0.5\ 10^{-6}$ m. Determine what is the minimum value needed for the kinetic energy of the protons in order to produce the reaction (proton + photon → proton + π), where the π particle has mass $m \simeq 0.14\ M$, the proton mass being $M \simeq 0.938$ GeV/c^2.

Answer: Let p and k be the momenta of the proton and of the photon respectively, $k = h/\lambda \simeq 2.48$ eV/c. The reaction can take place only if the invariant mass of the initial system is larger or equal to $(M + m)$: that is most easily seen in the center of mass frame, where the minimal energy condition corresponds to the two final particles being at rest. In particular, if E is the energy of the proton, we can write

$$(E + kc)^2 - (p - k)^2 c^2 \geq (M + m)^2 c^4, \quad \text{hence} \quad E + pc \geq \frac{mc^2}{kc}(M + m/2)c^2.$$

Taking into account that $mc^2 \simeq 0.135$ GeV and $kc \simeq 2.48$ eV we deduce that $E + pc \sim 5\ 10^7\ Mc^2$, so that the proton is ultra-relativistic and $E \simeq pc$. The minimal kinetic energy of the proton is therefore $p_{min}c \simeq (mc/k)(M + m/2)c^2/2 \simeq 2.66\ 10^7$ GeV.

1.33. A particle of mass M decays into two particles of masses m_1 and m_2. A detector reveals the energies and momenta of the outgoing particles to be $E_1 = 2.5$ GeV, $E_2 = 8$ GeV, $p_{1x} = 1$ GeV/c, $p_{1y} = 2.25$ GeV/c, $p_{2x} = 7.42$ GeV/c and $p_{2y} = 2.82$ GeV/c. Determine the masses of all involved particles, as well as the velocity $\boldsymbol{v}$ of the initial particle.

Answer: $M \simeq 3.69$ GeV/c^2, $m_1 \simeq 0.43$ GeV/c^2, $m_2 \simeq 1$ GeV/c^2, $v_x = 0.802\ c$, $v_y = 0.483\ c$.

1.34. A particle of mass $\mu = 0.14$ GeV/c^2 and momentum directed along the positive z axis, knocks into a particle at rest of mass M. The final state after the collision is made up of two particles of mass $m_1 = 0.5$ GeV/c^2 and $m_2 = 1.1$ GeV/c^2 respectively. The momenta of the two outgoing particles form an equal angle $\theta = 0.01$ rad with the z axis and have equal magnitude $p = 10^4$ GeV/c. What is the value of M?

Answer: Momentum conservation gives the momentum of the initial particle, $k = 2p\cos\theta$. The initial energy is therefore $E_{in} = \sqrt{\mu^2 c^4 + k^2 c^2} + M\ c^2$ and must be equal to the final energy $E_{fin} = \sqrt{m_1^2 c^4 + p^2 c^2} + \sqrt{m_2^2 c^4 + p^2 c^2}$, hence

$$Mc^2 = \sqrt{m_1^2 c^4 + p^2 c^2} + \sqrt{m_2^2 c^4 + p^2 c^2} - \sqrt{\mu^2 c^4 + 4p^2 \cos\theta^2 c^2}\,.$$

The very high value of p makes it sensible to apply the ultra-relativistic approximation,

$$Mc^2 \simeq pc\left(1 + \frac{m_1^2 c^2}{2p^2}\right) + pc\left(1 + \frac{m_2^2 c^2}{2p^2}\right) - 2pc\cos\theta\left(1 + \frac{\mu^2 c^2}{8p^2\cos\theta^2}\right)$$

$$\simeq pc\left(\theta^2 + (2m_1^2 + 2m_2^2 - \mu^2)c^2/(4p^2)\right) \simeq pc\ \theta^2 = 1 \text{ GeV}\,.$$

2

Introduction to Quantum Physics

The gestation of Quantum Physics has been very long and its phenomenological foundations were various. Historically the original idea came from the analysis of the black body spectrum. This is not surprising since the black body, in fact an oven in thermal equilibrium with the electromagnetic radiation, is a simple and fundamental system once the law of electrodynamics are established. As a matter of fact many properties of the spectrum can be deduced starting from the general laws of electrodynamics and thermodynamics; the crisis came from the violation of energy equipartition. This suggested to Planck the idea of *quantum*, from which everything originated. Of course a long sequence of different phenomenological evidences, first of all the photoelectric effect, the line nature of atomic spectra, the Compton effect and so on, gave a compelling evidence for the new theory.

Due to the particular limits of the present notes an exhaustive analysis of the whole phenomenology is impossible. Even a clear discussion of the black body problem needs an exceeding amount of space. Therefore we have chosen a particular line, putting major emphasis on the photoelectric effect and on the inadequacy of a classical approach based on Thomson's model of the atom, followed by Bohr's analysis of the quantized structure of Rutherford's atom and by the construction of Schrödinger's theory. This does not mean that we have completely overlooked the remaining phenomenology; we have just presented it in the light of the established quantum theory. Thus, for example, Chapter 3 ends with the analysis of the black body spectrum in the light of quantum theory.

2.1 The Photoelectric Effect

The photoelectric effect was discovered by H. Hertz in 1887. As sketched in Fig. 2.1, two electrodes are placed in a vacuum cell; one of them (C) is hit by monochromatic light of variable frequency, while the second (A) is set to a negative potential with respect to the first, as determined by a generator G and measured by a voltmeter V.

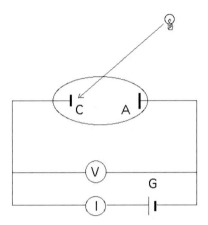

Fig. 2.1. A sketch of Hertz's photoelectric effect apparatus

By measuring the electric current going through the amperometer I, one observes that, if the light frequency is higher than a given threshold ν_V, determined by the potential difference V between the two electrodes, the amperometer reveals a flux of current i going from A to C which is proportional to the flux of luminous energy hitting C. The threshold ν_V is a linear function of the potential difference V

$$\nu_V = a + bV \ . \tag{2.1}$$

The reaction time of the apparatus to light is substantially determined by the (RC) time constant of the circuit and can be reduced down to values of the order of 10^{-8} s. The theoretical interpretation of this phenomenon remained an open issue for about 14 years because of the following reasons.

The current direction and the possibility to stop it by increasing the potential difference clearly show that the electric flux is made up of electrons pulled out from the atoms of electrode C by the luminous radiation.

A reasonable model for this process, which was inspired by Thomson's atomic model, assumed that electrons, which are particles of mass $m = 9 \ 10^{-31}$ Kg and electric charge $-e \simeq -1.6 \ 10^{-19}$ C, were elastically bound to atoms of size $R_A \sim 3 \ 10^{-10}$ m and subject to a viscous force of constant η. The value of η is determined as a function of the atomic relaxation time, $\tau = 2m/\eta$, that is the time needed by the atom to release its energy through radiation or collisions, which is of the order of $\tau = 10^{-8}$ s. Let us confine ourselves to considering the problem in one dimension and write the equation of motion for an electron

$$m\ddot{x} = -kx - \eta\dot{x} - eE\,, \tag{2.2}$$

where E is an applied electric field and k is determined on the basis of atomic frequencies. In particular we suppose the presence of many atoms with different frequencies continuously distributed around

$$\sqrt{\frac{k}{m}} = \omega_0 = 2\pi\nu_0 \sim 10^{15} \text{ s}^{-1}\,. \tag{2.3}$$

If we assume an oscillating electric field $E = E_0 \cos(\omega t)$ with $\omega \sim 10^{15} \text{ s}^{-1}$, corresponding to visible light, then a general solution to (2.2) is given by

$$x = x_0 \cos(\omega t + \phi) + A_1 e^{-\alpha_1 t} + A_2 e^{-\alpha_2 t}\,, \tag{2.4}$$

where the second and third term satisfy the homogeneous equation associated with (2.2), so that $\alpha_{1/2}$ are the solutions of the following equation

$$m\alpha^2 - \eta\alpha + k = 0\,,$$

$$\alpha = \frac{\eta \pm \sqrt{\eta^2 - 4km}}{2m} = \frac{1}{\tau} \pm \sqrt{\frac{1}{\tau^2} - \omega_0^2} \simeq \frac{1}{\tau} \pm i\,\omega_0\,, \tag{2.5}$$

where last approximation is due to the assumption $\tau \gg \omega_0^{-1}$.

Regarding the particular solution $x_0 \cos(\omega t + \phi)$, we obtain by substitution:

$$-m\omega^2 x_0 \cos(\omega t + \phi) = -kx_0 \cos(\omega t + \phi) + \eta\omega x_0 \sin(\omega t + \phi) - eE_0 \cos(\omega t) \tag{2.6}$$

hence

$$(k - m\omega^2)x_0 \left(\cos(\omega t)\cos\phi - \sin(\omega t)\sin\phi\right)$$
$$= \eta\omega x_0 \left(\sin(\omega t)\cos\phi + \cos(\omega t)\sin\phi\right) - eE_0 \cos(\omega t)$$

from which, by taking alternatively $\omega t = 0, \pi/2$, we obtain the following system

$$\left(m\left(\omega_0^2 - \omega^2\right)\cos\phi - \eta\omega\sin\phi\right)x_0 = -eE_0\,,$$
$$m\left(\omega_0^2 - \omega^2\right)x_0 \sin\phi + \eta\omega\,x_0\cos\phi = 0 \tag{2.7}$$

which can be solved for ϕ

$$\tan\phi = \frac{2\omega}{\tau\left(\omega^2 - \omega_0^2\right)}\,,$$

$$\cos\phi = \frac{\omega^2 - \omega_0^2}{\sqrt{\left(\omega_0^2 - \omega^2\right)^2 + \frac{4\omega^2}{\tau^2}}}\,, \qquad \sin\phi = \frac{(2\omega/\tau)}{\sqrt{\left(\omega_0^2 - \omega^2\right)^2 + \frac{4\omega^2}{\tau^2}}} \tag{2.8}$$

and finally for x_0, for which we obtain the well known resonant form

$$x_0 = \frac{eE_0/m}{\sqrt{\left(\omega_0^2 - \omega^2\right)^2 + \frac{4\omega^2}{\tau^2}}} \ . \tag{2.9}$$

To complete our computation we must determine A_1 and A_2. On the other hand, taking into account (2.5) and the fact that x is real, we can rewrite the general solution in the following equivalent form:

$$x = x_0 \cos(\omega t + \phi) + Ae^{-t/\tau} \cos(\omega_0 t + \phi_0) \ . \tag{2.10}$$

If we assume that the electron be initially at rest, we can determine A and ϕ_0 by taking $x = \dot{x} = 0$ for $t = 0$, i.e.

$$x_0 \cos \phi + A \cos \phi_0 = 0 \ , \tag{2.11}$$

$$x_0 \, \omega \sin \phi = -A \left(\frac{\cos \phi_0}{\tau} + \omega_0 \sin \phi_0 \right) \ , \tag{2.12}$$

hence in particular

$$\tan \phi_0 = \frac{\omega}{\omega_0} \tan \phi - \frac{1}{\omega_0 \tau} \ . \tag{2.13}$$

These equations give us enough information to discuss the photoelectric effect without explicitly substituting A in (2.10).

Indeed in our simplified model the effect, i.e. the liberation of the electron from the atomic bond, happens as the amplitude of the electron displacement x is greater than the atomic radius. In equation (2.10) x is the sum of two parts, the first corresponding to stationary oscillations, the second to a transient decaying with time constant τ. In principle, the maximum amplitude could take place during the transient or later: to decide which is the case we must compare the value of A with that of x_0. It is apparent from (2.11) that the magnitude of A is of the same order as x_0 unless $\cos \phi_0$ is much less than $\cos \phi$. On the other hand, equation (2.13) tells us that, if $\tan \phi_0$ is large, then $\tan \phi$ is large as well, since $(\omega_0 \tau)^{-1} \sim 10^{-7}$ and $\omega/\omega_0 \sim 1$. Therefore, the order of magnitude of the maximum displacement is given by x_0, and can be sensitive to the electric field frequency. That happens in the resonant regime, where ω differs from ω_0 by less than $2\sqrt{\omega/\tau}$.

Let us consider separately the generic case from the resonant one. In the first case the displacement is of the order of $eE_0/(\omega^2 m)$, since the square root of the denominator in (2.9) has the same order of magnitude as ω^2. In order to induce the photoelectric effect it is therefore necessary that

$$\frac{eE_0}{\omega^2 m} \sim R_A \ ,$$

from which we can compute the power density needed for the luminous beam which hits electrode C:

$$P = c\epsilon_0 E_0^2 \sim c\epsilon_0 \left(\frac{R_A \omega^2 m}{e} \right)^2 \ ,$$

where c is the speed of light and ϵ_0 is the vacuum dielectric constant. P comes out to be of the order of 10^{15} W/m^2, a power density which is difficult to realize in practice and which would anyway be enough to vaporize any kind of electrode. We must conclude that our model cannot explain the photoelectric effect if ω is far from resonance. Let us consider therefore the resonance case and set $\omega = \omega_0$. On the basis of (2.11), (2.13) and (2.9), that implies:

$$\phi = \phi_0 = \frac{\pi}{2} , \quad A = -x_0 ,$$

hence

$$x = \frac{-eE_0\tau}{2m\omega_0} \left(1 - e^{-t/\tau}\right) \sin(\omega_0 t) . \tag{2.14}$$

In order for the photoelectric effect to take place, the oscillation amplitude must be greater than the atomic radius:

$$\frac{eE_0\tau}{2m\omega_0} \left(1 - e^{-t/\tau}\right) \geq R_A .$$

That sets the threshold field to $2m\omega_0 R_A/(e\tau)$ and the power density of the beam to

$$P = c\epsilon_0 \left(\frac{4\omega_0 m R_A}{\tau e}\right)^2 \sim 100 \ \text{W/m}^2 ,$$

while the time required to reach the escape amplitude is of the order of τ.

In conclusion, our model predicts a threshold on the power of the beam, but not on its frequency, which however must be tuned to the resonance frequency: the photoelectric effect would cease both below and above the typical resonance frequencies of the atoms in the electrode. Moreover the expectation is that the electron does not gain any further appreciable energy from the electric field once it escapes the atomic bond: hence the emission from the electrode could be strong, but made up of electrons of energy equal to that gained during the last atomic oscillation. Equation (2.14) shows that, during the transient ($t << \tau$), the oscillation amplitude grows roughly by $eE_0/(m\omega_0^2)$ in one period, so that the energy of the escaped electron would be of the order of magnitude of $kR_A eE_0/(m\omega_0^2) = eE_0 R_A$, corresponding also to the energy acquired by the electron from the electric field E_0 when crossing the atom. It is easily computed that for a power density of the order of $10 - 100$ W/m^2, the electric field E_0 is roughly 100 V/m, so that the final kinetic energy of the electron would be 10^{-8} eV $\sim 10^{-27}$ J: this value is much smaller than the typical thermal energy at room temperature ($3kT/2 \sim 10^{-1}$ eV).

The prediction of the model is therefore in clear contradiction with the experimental results described above. In particular the very small energy of the emitted electrons implies that the electric current I should vanish even for small negative potential differences.

Einstein proposed a description of the effect based on the hypothesis that the energy be transferred from the luminous radiation to the electron in a single *elementary* (i.e. no further separable) process, instead than through a gradual excitation. Moreover he proposed that the transferred energy be equal to $h\nu = h\omega/(2\pi) \equiv \hbar\omega$, a quantity called *quantum* by Einstein himself. The constant h had been introduced by Planck several years before to describe the radiation emitted by an oven and its value is $6.63 \ 10^{-34}$ J s.

If the quantum of energy is enough for electron liberation, i.e. according to our model it is larger than $E_t \equiv kR_A^2/2 = \omega_0^2 R_A^2 m/2 \sim 10^{-19}$ J ~ 1 eV and the frequency exceeds $1.6 \ 10^{14}$ Hz (corresponding to ω in our model), then the electron is emitted keeping the energy exceeding the threshold in the form of kinetic energy. The number of emitted electrons, hence the intensity of the process, is proportional to the flux of luminous energy, i.e. to the number of quanta hitting the electrode.

Since $E = h\nu$ is the energy gained by the electron, which spends a part E_t to get free from the atom, the final electron kinetic energy is $T = h\nu - E_t$, so that the electric current can be interrupted by placing the second electrode at a negative potential

$$V = \frac{h\nu - E_t}{e},$$

thus reproducing (2.1).

The most important point in Einstein's proposal, which was already noticed by Planck, is that a physical system of typical frequency ν can exchange only quanta of energy equal to $h\nu$. The order of magnitude in the atomic case is $\omega \sim 10^{15}$ s^{-1}, hence $\hbar\omega \equiv (h/2\pi) \ \omega \sim 1$ eV.

2.2 Bohr's Quantum Theory

After the introduction of the concept of a quantum of energy, *quantum theory* was developed by N. Bohr in 1913 and then perfected by A. Sommerfeld in 1916: they gave a precise proposal for multi-periodic systems, i.e. systems which can be described in terms of periodic components.

The main purpose of their studies was that of explaining, in the framework of Rutherford's atomic model, the light spectra emitted by gasses (in particular mono-atomic ones) excited by electric discharges. The most simple and renowned case is that of the mono-atomic hydrogen gas (which can be prepared with some difficulties since hydrogen tends to form bi-atomic molecules). It has a discrete spectrum, i.e. the emitted frequencies can assume only some discrete values, in particular:

$$\nu_{n,m} = R\left(\frac{1}{n^2} - \frac{1}{m^2}\right) \tag{2.15}$$

for all possible positive integer pairs with $m > n$: this formula was first proposed by J. Balmer in 1885 for the case $n = 2$, $m \geq 3$, and then generalized by J. Rydberg in 1888 for all possible pairs (n, m). The emission is particularly strong for $m = n + 1$.

Rutherford had shown that the positive charge in an atom is localized in a practically point-like nucleus, which also contains most of the atomic mass. In particular the hydrogen atom can be described as a two-body system: a heavy and positively charged particle, which nowadays is called proton, bound by Coulomb forces to a light and negatively charged particle, the electron.

We will confine our discussion to the case of circular orbits of radius r, covered with uniform angular velocity ω, and will consider the proton as if it were infinitely heavy (its mass is about $2 \cdot 10^3$ times that of the electron). In this case we have

$$m\omega^2 r = \frac{e^2}{4\pi\epsilon_0 r^2} \, ,$$

where m is the electron mass. Hence the orbital frequencies, which in classical physics correspond to those of the emitted radiation, are continuously distributed as a function of the radius

$$\nu = \frac{\omega}{2\pi} = \frac{e}{\sqrt{16\pi^3\epsilon_0 m r^3}} \, , \tag{2.16}$$

this is in clear contradiction with (2.15). Based on Einstein's theory of the photoelectric effect, Bohr proposed to interpret (2.15) by assuming that only certain orbits be allowed in the atom, which are called *levels*, and that the frequency $\nu_{n,m}$ correspond to the transition from the m-th level to n-th one. In that case

$$h\nu_{n,m} = E_m - E_n \, , \tag{2.17}$$

where the atomic energies (which are negative since the atom is a bound system) would be given by

$$E_n = -\frac{hR}{n^2} \, . \tag{2.18}$$

Since, according to classical physics for the circular orbit case, the atomic energy is given by

$$E_{\mathrm{circ}} = -\frac{e^2}{8\pi\epsilon_0 r} \, ,$$

Bohr's hypothesis is equivalent to the assumption that the admitted orbital radii be

$$r_n = \frac{e^2 n^2}{8\pi\epsilon_0 h R} \, . \tag{2.19}$$

It is clear that Bohr's hypothesis seems simply aimed at reproducing the observed experimental data; it does not permit any particular further development, unless further conditions are introduced. The most natural, which is called *correspondence principle* is that the classical law, given in (2.16), be reproduced by (2.15) for large values of r, hence of n, and at least for the strongest emissions, i.e. those with $m = n + 1$, for which we can write

$$\nu_{n,n+1} = R\frac{2n+1}{n^2(n+1)^2} \to \frac{2R}{n^3} \, , \tag{2.20}$$

these frequencies should be identified in the above mentioned limit with what resulting from the combination of (2.16) and (2.19):

$$\nu = \frac{e}{\sqrt{16\pi^3\epsilon_0 m r_n^3}} = \frac{2^{\frac{5}{2}}\epsilon_0(hR)^{\frac{3}{2}}}{e^2\sqrt{mn^3}} . \tag{2.21}$$

By comparing last two equations we get the value of the coefficient R in (2.15), which is called *Rydberg constant*:

$$R = \frac{me^4}{8\epsilon_0^2 h^3}$$

and is in excellent agreement with experimental determinations. We have then the following *quantized* atomic energies

$$E_n = -\frac{me^4}{8\epsilon_0^2 h^2 n^2} , \qquad n = 1, 2, ...$$

while the quantized orbital radii are

$$r_n = \frac{\epsilon_0 h^2 n^2}{\pi m e^2} . \tag{2.22}$$

In order to give a numerical estimate of our results, it is convenient to introduce the ratio $e^2/(2\epsilon_0 hc) \equiv \alpha \simeq 1/137$, which is adimensional and is known as the *fine structure constant.*Constant!fine structure The energy of the state with $n = 1$, which is called the *fundamental state*, is

$$E_1 = -hR = -\frac{mc^2}{2}\alpha^2 ;$$

noticing that $mc^2 \sim 0.5\,\text{MeV}$, we have $E_1 \simeq -13\,\text{eV}$. The corresponding atomic radius is $r_1 \simeq 0.5\ 10^{-10}\,\text{m}$.

Notwithstanding the excellent agreement with experimental data, the starting hypothesis, to be identified with (2.18), looks still quite conditioned by the particular form of Balmer law given in (2.15). For that reason Bohr tried to identify a physical observable to be quantized according to a simpler and more fundamental law. He proceeded according to the idea that such observable should have the same dimensions of the Planck constant, i.e. those of an action, or equivalently of an angular momentum. In the particular case of quantized circular orbits this last quantity reads:

$$L = pr = m\omega r^2 = \frac{e}{\sqrt{4\pi\epsilon_0}}\sqrt{mr_n} = \frac{h}{2\pi}n \equiv n\hbar , \qquad n = 1, 2, ... \tag{2.23}$$

2.3 De Broglie's Interpretation

In this picture of partial results, even if quite convincing from the point of view of the phenomenological comparison, the real progress towards understanding quantum physics came as L. de Broglie suggested the existence of a *universal wave-like behavior* of material particles and of energy quanta associated to force fields. As we have seen in the case of electromagnetic waves when discussing the Doppler effect, a phase can always be associated with a wave-like process, which is variable both in space and in time (e.g. given by $2\pi \left(x/\lambda - \nu t \right)$ in the case of waves moving parallel to the x axis). The assumption that quanta can be interpreted as real particles and that Einstein's law $E = h\nu$ be universally valid, would correspond to identifying the wave phase with $2\pi \left(x/\lambda - Et/h \right)$. If we further assume the phase to be relativistically invariant, then it must be expressed in the form $\left(p\,x - E\,t \right)/\hbar$, where E and p are identified with relativistic energy and momentum, i.e. in the case of material particles:

$$E = \frac{mc^2}{\sqrt{\left(1 - \frac{v^2}{c^2}\right)}} \; , \qquad p = \frac{mv}{\sqrt{\left(1 - \frac{v^2}{c^2}\right)}} \; .$$

In order to simplify the discussion as much as possible, we will consider here and in most of the following a one dimensional motion (parallel to the x axis). In conclusion, by comparing last two expressions given for the phase, we obtain de Broglie's equation:

$$p = \frac{h}{\lambda} \; ,$$

which is complementary to Einstein's law, $E = h\nu$.

These formulae give an idea of the scale at which quantum effects are visible. For an electron having kinetic energy $E_k = 10^2$ eV $\simeq 1.6\ 10^{-17}$ J quantum effects show up at distances of the order of $\lambda = h/p = h/\sqrt{2mE_k} \sim 10^{-10}$ m, corresponding to atomic or slightly subatomic distances; that confirms the importance of quantum effects for electrons in condensed matter and in particular in solids, where typical energies are of the order of a few electron-Volts. For a gas of light atoms in equilibrium at temperature T, the kinetic energy predicted by equipartition theorem is $3kT/2$, where k is Boltzmann's constant. At a temperature $T = 300°$K (room temperature) the kinetic energy is roughly $2.5\ 10^{-2}$ eV, corresponding to wavelengths of about 10^{-10} m for atom masses of the order of 10^{-26} Kg. However at those distances the picture of a non-interacting (perfect) gas does not apply because of strong repulsive forces coming into play: in order to gain a factor ten on distances, it is necessary to reduce the temperature by a factor 100, going down to a few Kelvin degrees, at which quantum effects are manifest. For a macroscopic body of mass 1 Kg and kinetic energy 1 J quantum effects would show up at distances roughly equal to $3\ 10^{-34}$ m, hence completely negligible with respect to the thermal oscillation amplitudes of atoms, which are proportional to the square root of the absolute temperature, and are in particular of the order of a few nanometers at $T = 10^3$ °K, where the solid melts.

On the other hand, Einstein's formula gives us information about the scale of times involved in quantum processes, which is of the order of $h/\Delta E$, where

ΔE corresponds to the amount of exchanged energy. For $\Delta E \sim 1$ eV, times are roughly $4\ 10^{-15}$ s, while for thermal interactions at room temperature time intervals increase by a factor 40.

In conclusion, in the light of de Broglie's formula, quantum effects are not visible for macroscopic bodies and at macroscopic energies. For atoms in matter they show up after condensation or anyway at very low temperatures, while electrons in solids or in atoms are fully in the quantum regime.

In Rutherford's atomic model illustrated in previous Section, the circular motion of the electron around the proton must be associated, according to de Broglie, with a wave closed around the circular orbit. That resembles wave-like phenomena analogous to the oscillations of a ring-shaped elastic string or to air pressure waves in a toroidal reed pipe. That implies well tuned wavelengths, as in the case of musical instruments (which are not ring-shaped for obvious practical reasons). The need for tuned wavelength can be easily understood in the case of the toroidal reed pipe: a complete round of the ring must bring the phase back to its initial value, so that the total length of the pipe must be an integer multiple of the wavelength.

Taking into account previous equations regarding circular atomic orbits, we have the following electron wavelength:

$$\lambda = \frac{h}{p} = \frac{h}{e}\sqrt{\frac{4\pi\epsilon_0 r}{m}} \ ,$$

so that the tuning condition reads

$$2\pi r = n\lambda = \frac{nh}{e}\sqrt{\frac{4\pi\epsilon_0 r}{m}}$$

giving

$$r = \frac{n^2 h^2 \epsilon_0}{\pi e^2 m} \ ,$$

which confirms (2.22) and gives support to the picture proposed by Bohr and Sommerfeld. De Broglie's hypothesis, which was formulated in 1924, was confirmed in 1926 by Davisson and Gerner by measuring the intensity of an electron beam reflected by a nickel crystal. The apparatus used in the experiment is sketched in Fig. 2.2. The angular distribution of the electrons reflected in conditions of normal incidence shows a strongly anisotropic behavior with a marked dependence on the beam accelerating potential. In particular, an accelerating potential equal to 48 V leads to a quite pronounced peak at a reflection angle $\phi = 55.3°$. An analogous X-ray diffraction experiment permits to interpret the nickel crystal as an atomic lattice of spacing $0.215\ 10^{-9}$ m. The comparison between the angular distributions obtained for X-rays and for electrons shows relevant analogies, suggesting a diffractive interpretation also in the case of electrons. Bragg's law giving the n-th maximum in the diffraction figure is $d \sin \phi_n = n\lambda$.

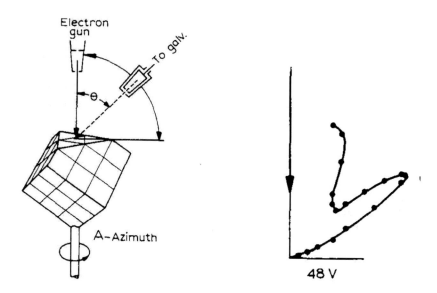

Fig. 2.2. A schematic description of Davisson-Gerner apparatus and a polar coordinate representation of the results obtained at 48 V electron energy, as they appear in Davisson's Noble Price Lecture, from *Nobel Lectures*, Physics 1922-1941 (Elsevier Publishing Company, Amsterdam, 1965)

For the peak corresponding to the principal maximum at 55.3° we have

$$d \, \sin \phi = \lambda \simeq 0.175 \; 10^{-9} \; \text{m} \, .$$

On the other hand the electrons in the beam have a kinetic energy

$$E_k \simeq 7.68 \; 10^{-18} \; \text{J} \, ,$$

hence a momentum $p \simeq 3.7 \; 10^{-24}$ N s, in excellent agreement with de Broglie's formula $p = h/\lambda$. In the following years analogous experiments were repeated using different kinds of material particles, in particular neutrons.

    Once established the wave-like behavior of propagating material particles, it must be clarified what is the physical quantity the phenomenon refers to, i.e. what is the physical meaning of the oscillating quantity (or quantities) usually called *wave function*, for which a *linear* propagating equation will be supposed, in analogy with mechanical or electromagnetic waves. It is known that, in the case of electromagnetic waves, the quantities measuring the amplitude are electric and magnetic fields. Our question regards exactly the analogous of those fields in the case of de Broglie's waves. The experiment by Davisson and Gerner gives an answer to this question. Indeed, as illustrated in Fig. 2.2, the detector reveals the presence of one or more electrons at a given angle; if we imagine to repeat the experiment several times, with a single electron in the beam at each time, and if we measure the frequency at which electrons are detected at the various angles, we get the *probability* of having the electron in a given site covered by the detector.

In the case of an optical measure, what is observed is the interference effect in the energy deposited on a plate; that is proportional to the square of the electric field on the plate. Notice that the linearity in the wave equation and the quadratic relation between the measured quantity and the wave amplitude are essential conditions for the existence of interference and diffractive phenomena. We must conclude that also in the case of material particles some *positive quadratic form* of de Broglie's wave function gives the probability of having the electron in a given point.

We have quite generically mentioned a quadratic form, since at the moment it is still not clear if the wave function has one or more components, i.e. if it corresponds to one or more real functions. By a positive quadratic form we mean a homogeneous second order polynomial in the wave function components, which is positive for real and non-vanishing values of its arguments. In the case of a single component, we can say without loss of generality that the probability density is the wave function squared, while in the case of two or more components it is always possible, by suitable linear transformations, to reduce the quadratic form to a sum of squares.

We are now going to show that the hypothesis of a single component must be discarded. Let us indicate by $\rho(\boldsymbol{r}, t)d^3r$ the probability of the particle being in a region of size d^3r around $\boldsymbol{r}$ at time t, and by $\psi(\boldsymbol{r}, t)$ the wave function, which for the moment is considered as a real valued function, defined so that

$$\rho(\boldsymbol{r}, t) = \psi^2(\boldsymbol{r}, t). \tag{2.24}$$

If Ω indicates the whole region accessible to the particle, the probability density must satisfy the natural constraint:

$$\int_\Omega d^3r \rho(\boldsymbol{r}, t) = 1, \tag{2.25}$$

which implies the condition:

$$\int_\Omega d^3r \, \dot{\rho}(\boldsymbol{r}, t) \equiv \int_\Omega d^3r \frac{\partial \rho(\boldsymbol{r}, t)}{\partial t} = 0. \tag{2.26}$$

This expresses the fact that, if the particle cannot escape Ω, the probability of finding it in that region must always correspond to certainty. This condition can be given in mathematical terms analogous to those used to express electric charge conservation: the charge contained in a given volume, i.e. the integral of the charge density, may change only if the charge flows through the boundary surface. The charge flux through the boundaries is expressed in terms of the current density flow and can be rewritten as the integral of the divergence of the current density itself by using Gauss–Green theorem

$$\int_{\Omega} \dot{\rho} = -\Phi_{\partial\Omega}(J) = -\int_{\Omega} \boldsymbol{\nabla} \cdot \boldsymbol{J} \ .$$

Finally, by reducing the equation from an integral form to a differential one, we can identify the temporal derivative of the charge density with the divergence of the current density. Based on this analogy, let us introduce the *probability current* density $\boldsymbol{J}$ and write

$$\dot{\rho}(\boldsymbol{r}, t) = -\frac{\partial J_x(\boldsymbol{r}, t)}{\partial x} - \frac{\partial J_y(\boldsymbol{r}, t)}{\partial y} - \frac{\partial J_z(\boldsymbol{r}, t)}{\partial z} \equiv -\boldsymbol{\nabla} \cdot \boldsymbol{J}(\boldsymbol{r}, t) \ . \qquad (2.27)$$

The conservation equation must be automatically satisfied as a consequence of the propagation equation of de Broglie's waves, which we write in the form:

$$\dot{\psi} = L\left(\psi, \boldsymbol{\nabla}\psi, \nabla^2\psi, ..\right) \ , \qquad (2.28)$$

where L indicates a generic linear function of ψ and its derivatives like:

$$L\left(\psi, \boldsymbol{\nabla}\psi, \nabla^2\psi, ..\right) = \alpha\psi + \beta\nabla^2\psi \ . \qquad (2.29)$$

Notice that if L were not linear the interference mechanism upon which quantization is founded would soon or later fail. Furthermore we assume invariance under the reflection of coordinates, so that terms proportional to first derivatives are excluded.

From equation (2.24) we have $\dot{\rho} = 2\psi\dot{\psi}$, which can be rewritten, using (2.28), as:

$$\dot{\rho} = 2\psi L\left(\psi, \boldsymbol{\nabla}\psi, \nabla^2\psi, ..\right) \ . \qquad (2.30)$$

The right-hand side of last equation must be identified with $-\boldsymbol{\nabla}\cdot\boldsymbol{J}(\boldsymbol{r}, t)$. Moreover $\boldsymbol{J}$ must necessarily be a bilinear function of ψ and its derivatives exactly like $\dot{\rho}$. Therefore, since $\boldsymbol{J}$ is a vector-like quantity, it must be expressible as

$$\boldsymbol{J} = c\ \psi\boldsymbol{\nabla}\psi + d\ \boldsymbol{\nabla}\psi\nabla^2\psi + \dots$$

from which it appears that $\boldsymbol{\nabla} \cdot \boldsymbol{J}(\boldsymbol{r}, t)$ must necessarily contain bilinear terms in which both functions are derived, like $\boldsymbol{\nabla}\psi \cdot \boldsymbol{\nabla}\psi$: however such terms are clearly missing in (2.30).

We come to the conclusion that the description of de Broglie's waves requires at least two wave functions ψ_1 and ψ_2, defined so that $\rho = \psi_1^2 + \psi_2^2$. In an analogous way we can introduce the complex valued function:

$$\psi = \psi_1 + i\psi_2 \ , \qquad (2.31)$$

defined so that

$$\rho = |\psi|^2 \; ; \qquad\qquad (2.32)$$

this choice implies:

$$\dot{\rho} = \psi^* \dot{\psi} + \psi \dot{\psi}^* \, .$$

If we assume, for instance, the wave equation corresponding to (2.29):

$$\dot{\psi} = \alpha\psi + \beta\nabla^2\psi \, , \qquad\qquad (2.33)$$

we obtain:

$$\dot{\rho} = \psi^* \left(\alpha\psi + \beta\nabla^2\psi \right) + \psi \left(\alpha^*\psi^* + \beta^*\nabla^2\psi^* \right) \, .$$

If we also assume that the current probability density be

$$\boldsymbol{J} = \mathrm{i}k \left(\psi^* \boldsymbol{\nabla}\psi - \psi\boldsymbol{\nabla}\psi^* \right) \, , \qquad\qquad (2.34)$$

with k real so as to make $\boldsymbol{J}$ real as well, we easily derive

$$\boldsymbol{\nabla} \cdot \boldsymbol{J} = \mathrm{i}k \left(\psi^*\nabla^2\psi - \psi\nabla^2\psi^* \right) \, .$$

It can be easily verified that the continuity equation (2.27) is satisfied if

$$\alpha + \alpha^* = 0 \, , \qquad \beta = -\mathrm{i}k \, . \qquad\qquad (2.35)$$

It is of great physical interest to consider the case in which the wave function has more than two real components. In particular, the wave function of electrons has four components or, equivalently, two complex components. In general, the multiplicity of the complex components is linked to the existence of an intrinsic angular momentum, which is called *spin*. The various complex components are associated with the different possible spin orientations. In the case of particles with non-vanishing mass, the number of components is $2S+1$, where S is the spin of the particle. In the case of the electron, $S = 1/2$.

For several particles, as for the electron, spin is associated with a magnetic moment which is inherent to the particle: it behaves as a microscopic magnet with various possible orientations, corresponding to those of the spin, which can be selected by placing the particle in a non-uniform magnetic field and measuring the force acting on the particle.

2.4 Schrödinger's Equation

The simplest case to which our considerations can be applied is that of a non-relativistic free particle of mass m. To simplify notations and computations, we will confine ourselves to a one-dimensional motion, parallel for instance, to the x axis; if the particle is not free, forces will be parallel to the same axis as well. The obtained results will be extensible to three dimensions by exploiting the vector formalism. In practice, we will sistematically replace $\boldsymbol{\nabla}$ by its component $\nabla_x = \partial/\partial x \equiv \partial_x$ and the Laplacian operator $\nabla^2 =$

$\partial^2/\partial x^2 + \partial^2/\partial y^2 + \partial^2/\partial z^2$ by $\partial^2/\partial x^2 \equiv \partial_x^2$; the probability current density $\boldsymbol{J}$ will be replaced by J_x (J) as well. The inverse replacement will suffice to get back to three dimensions.

The energy of a non-relativistic free particle is

$$E = c\sqrt{m^2c^2 + p^2} \simeq mc^2 + \frac{p^2}{2m} + O\left(\frac{p^4}{m^3c^2}\right),$$

where we have explicitly declared our intention to neglect terms of the order of $p^4/(m^3c^2)$. Assuming de Broglie's interpretation, we write the wave function:

$$\psi_P(x,t) \sim e^{2\pi i(x/\lambda - \nu t)} = e^{i(px - Et)/\hbar} \tag{2.36}$$

(we are considering a motion in the positive x direction). Our choice implies the following wave equation

$$\dot{\psi}_P = -\frac{iE}{\hbar}\psi_P = -\frac{i}{\hbar}\left(mc^2 + \frac{1}{2m}p^2\right)\psi_P. \tag{2.37}$$

We have also

$$\partial_x \psi_P = \frac{i}{\hbar}p\psi_P, \tag{2.38}$$

from which we deduce

$$i\hbar\dot{\psi}_P = mc^2\psi_P - \frac{\hbar^2}{2m}\partial_x^2\psi_P. \tag{2.39}$$

Our construction can be simplified by multiplying the initial wave function by the phase factor $e^{imc^2t/\hbar}$, i.e. defining

$$\psi \equiv e^{imc^2t/\hbar}\psi_P \sim \exp\left(\frac{i}{\hbar}\left(px - \frac{p^2}{2m}t\right)\right). \tag{2.40}$$

Since the dependence on x is unchanged, ψ still satisfies (2.38) and has the same probabilistic interpretation as ψ_P. Indeed both ρ and J are unchanged. The wave equation instead changes:

$$i\hbar\dot{\psi} = -\frac{\hbar^2}{2m}\partial_x^2\psi \equiv T\psi. \tag{2.41}$$

This is the *Schrödinger equation* for a free (non-relativistic) particle, in which the right-hand side has a natural interpretation in terms of the particle energy, which in the free case is only of kinetic type.

In the case of particles under the influence of a force field corresponding to a potential energy $V(x)$, the equation can be generalized by adding $V(x)$ to the kinetic energy :

$$i\hbar\dot{\psi} = -\frac{\hbar^2}{2m}\partial_x^2\psi + V(x)\psi. \tag{2.42}$$

This is the one-dimensional Schrödinger equation that we shall apply to various cases of physical interest.

Equations (2.34) and (2.35) show that the probability density current does not depend on V and is given by:

$$J = -\frac{i\hbar}{2m}\left(\psi^* \partial_x \psi - \psi \partial_x \psi^*\right) . \tag{2.43}$$

Going back to the free case and considering the *plane* wave function given in (2.36), it is interesting to notice that the corresponding probability density, $\rho = |\psi|^2$, is a constant function. This result is paradoxical since, by reducing (2.25) to one dimension, we obtain

$$\int_{-\infty}^{\infty} dx \; \rho(x,t) = \int_{-\infty}^{\infty} dx \; |\psi(x,t)|^2 = 1 , \tag{2.44}$$

which cannot be satisfied in the examined case since the integral of a constant function is divergent. We must conclude that our interpretation excludes the possibility that a particle have a well defined momentum.

We are left with the hope that this difficulty may be overcome by admitting some (small) uncertainty on the knowledge of momentum. This possibility can be easily analyzed thanks to the linearity of Schrödinger equation. Indeed equation (2.41) admits other different solutions besides the simple plane wave, in particular the *wave packet* solution, which is constructed as a linear superposition of many plane waves according to the following integral:

$$\int_{-\infty}^{\infty} dp \; \tilde{\psi}(p) \exp\left(\frac{i}{\hbar}\left(px - \frac{p^2}{2m}t\right)\right) .$$

The squared modulus of the superposition coefficients, $|\tilde{\psi}(p)|^2$, can be naturally interpreted, apart from a normalization constant, as the probability density in terms of momentum, exactly in the same way as $\rho(x)$ is interpreted as a probability density in terms of position.

Let us choose in particular a Gaussian distribution:

$$\tilde{\psi}(p) \sim e^{-(p-p_0)^2/(4\Delta^2)} , \tag{2.45}$$

corresponding to

$$\psi_\Delta(x,t) = k \int_{-\infty}^{\infty} dp \; e^{-(p-p_0)^2/(4\Delta^2)} \; e^{i\left(px - p^2 t/2m\right)/\hbar} . \tag{2.46}$$

where k must be determined in such a way that $\int_{-\infty}^{\infty} dx |\psi_\Delta(x,t)|^2 = 1$.

The integral in (2.46) can be computed by recalling that, if α is a complex number with positive real part $(Re\,(\alpha) > 0)$, then

$$\int_{-\infty}^{\infty} dp \; e^{-\alpha p^2} = \sqrt{\frac{\pi}{\alpha}}$$

and that the Riemann integral measure dp is left invariant by translations in the complex plane,

$$\int_{-\infty}^{\infty} dp \; e^{-\alpha p^2} \equiv \int_{-\infty}^{\infty} d(p+\gamma) \; e^{-\alpha(p+\gamma)^2}$$

$$= \int_{-\infty}^{\infty} dp \; e^{-\alpha(p+\gamma)^2} = e^{-\alpha\gamma^2} \int_{-\infty}^{\infty} dp \; e^{-\alpha p^2} e^{-2\alpha\gamma p},$$

for every complex number γ. Therefore we have

$$\int_{-\infty}^{\infty} dp \; e^{-\alpha p^2} e^{\beta p} = \sqrt{\frac{\pi}{\alpha}} e^{\beta^2/4\alpha} . \tag{2.47}$$

Developing (2.46) with the help of (2.47) we can write

$$\psi_\Delta(x,t) = k e^{-\frac{p_0^2}{4\Delta^2}} \int_{-\infty}^{\infty} dp \; e^{-\left[\frac{1}{4\Delta^2} + \frac{it}{2m\hbar}\right]p^2} \; e^{\left[\frac{p_0}{2\Delta^2} + \frac{ix}{\hbar}\right]p}$$

$$= k \sqrt{\frac{\pi}{\frac{1}{4\Delta^2} + \frac{it}{2m\hbar}}} \; \exp\left(\frac{\left[\frac{p_0}{2\Delta^2} + \frac{ix}{\hbar}\right]^2}{\frac{1}{\Delta^2} + \frac{2it}{m\hbar}} - \frac{p_0^2}{4\Delta^2}\right) . \tag{2.48}$$

We are interested in particular in the x dependence of the probability density $\rho(x)$: that is solely related to the real part of the exponent of the rightmost term in (2.48), which can be expanded as follows:

$$\frac{\frac{p_0^2}{4\Delta^4} + \frac{ip_0 x}{\Delta^2 \hbar} - \frac{x^2}{\hbar^2}}{\frac{1}{\Delta^2} + \frac{2it}{m\hbar}} - \frac{p_0^2}{4\Delta^2} = \frac{p_0^2}{4\Delta^2} \frac{\frac{4t^2\Delta^4}{m^2\hbar^2} + \frac{2it\Delta^2}{m\hbar}}{1 + \frac{4t^2\Delta^4}{m^2\hbar^2}} - \left(\frac{\Delta^2 x^2}{\hbar^2} - \frac{ip_0 x}{\hbar}\right)\frac{1 - \frac{2it\Delta^2}{m\hbar}}{1 + \frac{4t^2\Delta^4}{m^2\hbar^2}}$$

the real part being

$$-\frac{\Delta^2\left(x - \frac{p_0 t}{m}\right)^2}{\hbar^2\left(1 + \frac{4t^2\Delta^4}{m^2\hbar^2}\right)} \equiv -\frac{\Delta^2\left(x - v_0 t\right)^2}{\hbar^2\left(1 + \frac{4t^2\Delta^4}{m^2\hbar^2}\right)} .$$

Since p_0 is clearly the average momentum of the particle, we have introduced the corresponding average velocity $v_0 = p_0/m$. Recalling the definition of ρ as well as its normalization constraint, we finally find

$$\rho(x,t) = \frac{\Delta}{\hbar}\sqrt{\frac{2}{\pi\left(1 + \frac{4t^2\Delta^4}{m^2\hbar^2}\right)}} \exp\left(-\frac{2\Delta^2}{\hbar^2}\frac{\left(x - v_0 t\right)^2}{1 + \frac{4t^2\Delta^4}{m^2\hbar^2}}\right) , \tag{2.49}$$

while the probability distribution in terms of momentum reads

$$\tilde{\rho}(p) = \frac{1}{\sqrt{2\pi}\Delta} e^{-(p-p_0)^2/(2\Delta^2)} . \tag{2.50}$$

Given a Gaussian distribution $\rho(x) = 1/(\sqrt{2\pi}\sigma)e^{-(x-x_0)^2/(2\sigma^2)}$, it is a well known fact, which anyway can be easily derived from previous formulae, that the mean value $\bar{x}$ is x_0 while the mean quadratic deviation $\overline{(x - \bar{x})^2}$ is equal to σ^2. Hence, in the examined case, we have an average position $\bar{x} = v_0 t$ with a mean quadratic deviation equal to $\hbar^2/(4\Delta^2) + t^2\Delta^2/m^2$, while the average momentum is p_0 with a mean quadratic deviation Δ^2. The mean values represent the kinematic variables of a free particle, while the mean quadratic deviations are roughly inversely proportional to each other: if we improve the definition of one observable, the other becomes automatically less defined.

The distributions given in (2.49) and (2.50), even if derived in the context of a particular example, permit to reach important general conclusions which, for the sake of clarity, are listed in the following as distinct points.

2.4.1 The Uncertainty Principle

While the mean quadratic deviation relative to the momentum distribution

$$\overline{(p - \bar{p})^2} = \Delta^2$$

has been fixed a-priori by choosing $\tilde{\psi}(p)$ and is independent of time, thus confirming that momentum is a constant of motion for a free particle, that relative to the position

$$\overline{(x - \bar{x})^2} = \left(1 + \frac{4t^2\Delta^4}{m^2\hbar^2}\right)\frac{\hbar^2}{4\Delta^2}$$

does not contain further free parameters and does depend on time. Indeed, Δ_x grows significantly for $2t\Delta^2/(m\hbar) > 1$, hence for times greater than $t_s = m\hbar/(2\Delta^2)$. Notice that t_s is nothing but the time needed for a particle of momentum Δ to cover a distance $\hbar/(2\Delta)$, therefore this spreading has a natural interpretation also from a classical point of view: a set of independent particles having momenta distributed according to a width Δ_p, spreads with velocity $\Delta_p/m = v_s$; if the particles are statistically distributed in a region of size initially equal to Δ_x, the same size will grow significantly after times of the order of Δ_x/v_s.

What is new in our results is, first of all, that they refer to a single particle, meaning that uncertainties in position and momentum are not avoidable; secondly, these uncertainties are strictly interrelated. Without considering the spreading in time, it is evident that the uncertainty in one variable can be diminished only as the other uncertainty grows. Indeed, Δ can be eliminated from our equations by writing the inequality:

$$\Delta_x \Delta_p \equiv \sqrt{\overline{(x - \bar{x})^2} \; \overline{(p - \bar{p})^2}} \geq \frac{\hbar}{2}, \qquad (2.51)$$

which is known as the *Heinsenberg uncertainty principle* and can be shown to be valid for any kind of wave packet. The case of a real Gaussian packet corresponds to the minimal possible value $\Delta_x \Delta_p = \hbar/2$.

From a phenomenological point of view this principle originates from the universality of diffractive phenomena. Indeed diffractive effects are those which prevent the possibility of a simultaneous measurement of position and momentum with arbitrarily good precision for both quantities. Let us consider for instance the case in which the measurement is performed through optical instruments; in order to improve the resolution it is necessary to make use of radiation of shorter wavelength, thus increasing the momenta of photons, which hitting the object under observation change its momentum in an unpredictable way. If instead position is determined through mechanical instruments, like slits, then the uncertainty in momentum is caused by diffractive phenomena.

It is important to evaluate the order of magnitude of quantum uncertainty in cases of practical interest. Let us consider for instance a beam of electrons emitted by a cathode at a temperature $T = 1000°$K and accelerated through a potential difference equal to 10^4 V. The order of magnitude of the kinetic energy uncertainty Δ_E is kT, where $k = 1.381 \ 10^{-23}$ J/$°$K is the Boltzmann constant (alternatively one can use $k = 8.617 \ 10^{-5}$ eV/$°$K). Therefore $\Delta_E = 1.38 \ 10^{-20}$ J while $E = 1.6 \ 10^{-15}$ J, corresponding to a quite precise determination of the beam energy $(\Delta_E/E \sim 10^{-5})$. We can easily compute the momentum uncertainty by using error propagation $(\Delta_p/p = \frac{1}{2}\Delta_E/E)$ and computing $p = \sqrt{2m_e E} = 5.6 \ 10^{-23}$ N s; we thus obtain $\Delta_p = 2.8 \ 10^{-28}$ N s, hence, making use of (2.51), $\Delta_x \geq 2 \ 10^{-7}$ m. It is clear that the uncertainty principle does not place significant constraints in the case of particle beams.

A macroscopic body of mass $M = 1$ Kg placed at room temperature $(T \simeq 300°$K) has an average thermal momentum , caused by collisions with air molecules, which is equal to $\Delta_p \sim \sqrt{2M \ 3kT/2} \simeq 9 \ 10^{-11}$ N s, so that the minimal quantum uncertainty on its position is $\Delta_x \sim 10^{-24}$ m, hence not appreciable.

The uncertainty principle is instead quite relevant at the atomic level, where it is the stabilizing mechanism which prevents the electron from collapsing onto the nucleus. We can think of the electron orbital radius as a rough estimate of its position uncertainty $(\Delta_x \sim r)$ and evaluate the kinetic energy deriving from the momentum uncertainty; we have $E_k \sim \Delta_p^2/(2m) \sim \hbar^2/(2mr^2)$. Taking into account the binding Coulomb energy, the total energy is

$$E(r) \sim \frac{\hbar^2}{2mr^2} - \frac{e^2}{4\pi\epsilon_0 r} \, .$$

We infer that the system is stable, since the total energy $E(r)$ has an absolute minimum. The stable radius r_m corresponding to this minimum can be computed through the equation

$$\frac{e^2}{4\pi\epsilon_0 r_m^2} - \frac{\hbar^2}{mr_m^3} = 0 \, ,$$

hence

$$r_m \sim \frac{4\pi\epsilon_0 \hbar^2}{me^2} ,$$

which nicely reproduce the value of the atomic radius for the fundamental level in Bohr's model, see (2.22).

2.4.2 The Speed of Waves

It is known that electromagnetic waves move without distortion at a speed $c = 1/\sqrt{\epsilon_0 \mu_0}$ and that, for a harmonic wave, c is given by the wavelength multiplied by the frequency.

In the case of de Broglie's waves introduced in (2.40), we have $\nu = p^2/(2mh)$ and $\lambda = h/p$; therefore the velocity of harmonic waves is given by $v_F \equiv \lambda\nu = p/(2m)$. If we consider instead the wave packet given in (2.48) and its corresponding probability density given in (2.49), we clearly see that it moves with a velocity $v_G \equiv p_0/m$, which is equal to the classical velocity of a particle with momentum p_0. We have used different symbols to distinguish the velocity of plane waves v_F, which is called *phase velocity*, from v_G, which is the speed of the packet and is called *group velocity*. Previous equations lead to the result that, contrary to what happens for electromagnetic waves propagating in vacuum, the two velocities are different for de Broglie's waves, and in particular the group velocity does not coincide with the average value of the phase velocities of the different plane waves making up the packet. Moreover, the phase velocity depends on the wavelength ($v_F = h/(2m\lambda)$). The relation between frequency and wavelength is given by $\nu = c/\lambda$ for electromagnetic waves, while for de Broglie's waves it is $\nu = h/(2m\lambda^2)$.

There is a very large number of examples of wave-like propagation in physics: electromagnetic waves, elastic waves, gravity waves in liquids and several other ones. In each case the frequency presents a characteristic dependence on the wavelength, $\nu(\lambda)$. Considering as above the propagation of gaussian wave packets, it is always possible to define the phase velocity, $v_F = \lambda \, \nu(\lambda)$, and the group velocity, which in general is defined by the relation:

$$v_G = -\lambda^2 \frac{d\nu(\lambda)}{d\lambda} . \tag{2.52}$$

Last equation can be verified by considering that, for a generic dependence of the wave phase on the wave number $\exp(ikx - i\omega(k)t)$ and for a generic wave packet described by superposition coefficients strongly peaked around a given value $k = k_0$, the resulting wave function

$$\psi(x) \propto \int_{-\infty}^{\infty} dk \, f(k - k_0) \, e^{i(kx - \omega(k)t)} .$$

will be peaked around an x_0 such that the phase factor is stationary, hence almost constant, for $k \sim k_0$, leading to $x_0 \sim \omega'(k_0)t$.

In the case of de Broglie's waves (2.52) reproduces the result found previously. Media where the frequency is inversely proportional to the wavelength, as for electromagnetic waves in vacuum, are called *non-dispersive media*, and in that case the two velocities coincide.

It may be interesting to notice that, if we adopt the relativistic form for the plane wave, we have $\nu(\lambda) = \sqrt{m^2c^4/h^2 + c^2/\lambda^2}$, hence

$$v_F = \lambda\sqrt{\frac{m^2c^4}{h^2} + \frac{c^2}{\lambda^2}} = \frac{E}{p} > c\,,$$

$$v_G = \frac{c^2}{\lambda}\left(\frac{m^2c^4}{h^2} + \frac{c^2}{\lambda^2}\right)^{-1/2} = \frac{pc^2}{E} < c\,.$$

In particular v_G, which describes the motion of wave packets, satisfies the constraint of being less than c and coincides with the relativistic expression for the speed of a particle in terms of momentum and energy given in Chapter 1.

2.4.3 The Collective Interpretation of de Broglie's Waves

The description of single particles as wave packets is at the basis of a rigorous formulation of Schrödinger's theory. There is however an alternative interpretation of the wave function, which is of much simpler use and can be particularly useful to describe average properties, like a particle flow in the free case.

Let us consider the plane wave in (2.40): $\psi = \exp\left(i\left(p\,x - p^2t/(2m)\right)/\hbar\right)$ and compute the corresponding current density J:

$$J = -\frac{i\hbar}{2m}\left(\psi^*\partial_x\psi - \psi\partial_x\psi^*\right) = -\frac{i\hbar}{2m}\left(\psi^*\frac{ip}{\hbar}\psi - \psi\frac{-ip}{\hbar}\psi^*\right) = \frac{p}{m}\,, \quad (2.53)$$

while $\rho = \psi^*\psi = 1$. On the other hand we notice that given a distribution of classical particles with density ρ and moving with velocity v, the corresponding current density is $J = \rho v$.

That suggests to go beyond the problem of normalizing the probability distribution in (2.44), relating instead the wave function in (2.40) not to a single particle, as we have done till now, but to a stationary flux of independent particles, which are uniformly distributed with unitary density and move with the same velocity v.

It should be clear that in this way we are a priori giving up the idea of particle localization, however we obtain in a much simpler way information about the group velocity and the flux. We will thus be able, in the following Chapter, to easily and clearly interpret the effects of a potential barrier on a particle flux.

2.5 The Potential Barrier

The most interesting physical situation is that in which particles are not free, but subject to forces corresponding to a potential energy $V(x)$. In these conditions the Schrödinger equation in the form given in (2.42) has to be used. Since the equation is linear, the study can be limited, without loss of generality, to solutions which are periodic in time, like:

$$\psi(x,t) = e^{-iEt/\hbar}\psi_E(x).$$
(2.54)

Indeed the general time dependent solution can always be decomposed in periodic components through a Fourier expansion, so that its knowledge is equivalent to that of $\psi_E(x)$ plus the expansion coefficients.

Furthermore, according to the collective interpretation of de Broglie waves presented in last Section, the wave function in (2.54) describes either a stationary flow or a stationary state of particles. In particular we shall begin studying a stationary flow hitting a potential barrier.

The function $\psi_E(x)$ is a solution of the equation obtained by replacing (2.54) into (2.42), i.e.

$$i\hbar\ \partial_t e^{-iEt/\hbar}\psi_E(x) = E e^{-iEt/\hbar}\psi_E(x) = e^{-iEt/\hbar}\left[-\frac{\hbar^2}{2m}\partial_x^2\psi_E + V(x)\psi_E\right]$$
(2.55)

hence

$$E\psi_E(x) = -\frac{\hbar^2}{2m}\partial_x^2\psi_E(x) + V(x)\psi_E(x),$$
(2.56)

which is known as the *time-independent* or *stationary* schrödinger equation.

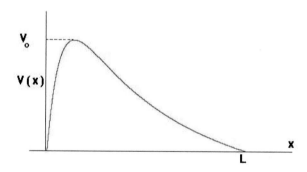

Fig. 2.3. A typical example of a potential barrier, referring in particular to that due to Coulomb repulsion that will be used when discussing Gamow's theory of nuclear α-emission

We will consider at first the case of a *potential barrier*, in which $V(x)$ vanishes for $x < 0$ and $x > L$, and is positive in the segment $[0, L]$, as shown in Fig. 2.3. A flux of classical particles hitting the barrier from the left will experience slowing forces as $x > 0$. If the starting kinetic energy, corresponding in this case to the total energy E in (2.56), is greater than the barrier height V_0, the particles will reach the point where V has a maximum, being accelerated from there forward till they pass point $x = L$, where the motion gets free again. Therefore the flux is completely transmitted, the effect of the barrier being simply a slowing down in the segment $[0, L]$. If instead the kinetic energy is less than V_0, the particles will stop before they reach the point where V has a maximum, reversing their motion afterwards: the flux is completely reflected in this case. Quantum Mechanics gives a completely different result.

In order to analyze the differences from a qualitative point of view, it is convenient to choose a barrier which makes the solution of (2.56) easier: that is the case of a potential which is piecewise constant, like the square barrier depicted on the side. The choice is motivated by the fact that, if V is constant, then (2.56) can be rewritten as follows:

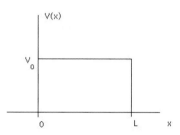

$$\partial_x^2 \psi_E(x) + \frac{2m}{\hbar^2}(E - V)\psi_E(x) = 0 \,, \tag{2.57}$$

and has the general solution:

$$\psi_E(x) = a_+ \exp\left(i\frac{\sqrt{2m(E - V)}}{\hbar}x\right) + a_- \exp\left(-i\frac{\sqrt{2m(E - V)}}{\hbar}x\right) \,, \tag{2.58}$$

if $E > V$, while

$$\psi_E(x) = a_+ \exp\left(\frac{\sqrt{2m(V - E)}}{\hbar}x\right) + a_- \exp\left(-\frac{\sqrt{2m(V - E)}}{\hbar}x\right) \,, \tag{2.59}$$

in the opposite case. The problem is then to establish how the solution found in a definite region can be connected to those found in the nearby regions. In order to solve this kind of problem we must be able to manage differential equations in presence of discontinuities in their coefficients, and that requires a brief *mathematical interlude*.

2.5.1 Mathematical Interlude: Differential Equations with Discontinuous Coefficients

Differential equations with discontinuous coefficients can be treated by smoothing the discontinuities, then solving the equations in terms of functions which are derivable several times, and finally reproducing the correct solutions in presence of discontinuities through a limit process. In order to do so, let us introduce the function $\varphi_\epsilon(x)$, which is defined as

$$\varphi_\epsilon(x) = 0 \quad \text{if} \quad |x| > \epsilon \,,$$

$$\varphi_\epsilon(x) = \frac{\epsilon^2 + x^2}{2\left(\epsilon^2 - x^2\right)^2} \frac{1}{\cosh^2\left(x/(\epsilon^2 - x^2)\right)} \quad \text{if} \quad |x| < \epsilon \,.$$

This function, as well as all of its derivatives, is continuous and it can be easily shown that

$$\int_{-\infty}^{\infty} \varphi_\epsilon(x) dx = 1 \,.$$

Based on this property we conclude that if $f(x)$ is locally integrable, i.e. if it admits at most isolated singularities where the function may diverge with a degree less than one, like for instance $1/|x|^{1-\delta}$ when $\delta > 0$, then the integral

$$\int_{-\infty}^{\infty} \varphi_\epsilon(x - y)f(y) dy \equiv f_\epsilon(x)$$

defines a function which can be derived in x an infinite number of times; the derivatives of f_ϵ tend to those of f in the limit $\epsilon \to 0$ and in all points where the latter are defined. We have in particular, by part integration,

$$\frac{d^n}{dx^n} f_\epsilon(x) = \int_{-\infty}^{\infty} \varphi_\epsilon(x - y) \frac{d^n}{dy^n} f(y) dy \,. \qquad (2.60)$$

f_ϵ is called *regularized function*. If for instance we consider the case in which f is the step function in the origin, i.e. $f(x) = 0$ for $x < 0$ and $f(x) = 1$ for $x > 0$, we have for $f_\epsilon(x)$, $\partial_x f_\epsilon(x) = f_\epsilon'(x)$ and $\partial_x^2 f_\epsilon(x) = f_\epsilon''(x)$ the behaviours showed in the respective order on the side. Notice in particular that since

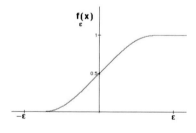

$$f_\epsilon(x) = \int_0^{\infty} \varphi_\epsilon(x - y) dy = \int_{-\infty}^{x} \varphi_\epsilon(z) dz$$

we have $\partial_x f_\epsilon(x) = \varphi_\epsilon(x)$. By looking at the three figures it is clear that $f_\epsilon(x)$ continuously interpolates between the two values, zero and one, which the function assumes respectively to the left of $-\epsilon$ and to the right of ϵ, staying less than 1 for every value of x. It is important to notice that instead the second figure, showing $\partial_x f_\epsilon(x)$, i.e. $\varphi_\epsilon(x)$, has a maximum of height proportional to $1/\epsilon^2$, hence diverging as $\epsilon \to 0$.

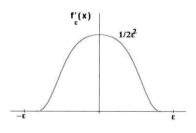

The third figure, showing the second derivative $\partial_x^2 f_\epsilon(x)$, has an oscillation of amplitude proportional to $1/\epsilon^4$ around the discontinuity point. Since, for small ϵ, the regularized function depends, close to the discontinuity, on the nearby values of the original function, it is clear that the qualitative behaviors showed in the figures are valid, close to discontinuities of the first kind (i.e. where the function itself has a discontinuous gap), for every starting function f.

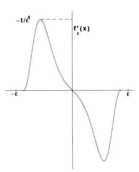

Let us now consider (2.57) close to a discontinuity point of the first kind (step function) for V, and suppose we regularize both terms on the left hand side. Assuming that the wave function do not present discontinuities worse than first kind, the second term in the equation may present only steps so that, once regularized, it is limited independently of ϵ. However the first term may present oscillations of amplitude $\sim 1/\epsilon^4$ if ψ_E has a first kind discontinuity, or a peak of height $\sim \pm 1/\epsilon^2$ if ψ_E is continuous but its first derivative has such discontinuity: in each case the modulus of the first regularized term would diverge faster than the second in the limit $\epsilon \to 0$. That shows that in presence of a first kind discontinuity in V, both the wave function ψ_E and its derivative must be continuous.

In order to simply deal with barriers of length L much smaller than the typical wavelengths of the problem, it is useful to introduce infinitely thin barriers: that can be done by choosing a potential energy which, once regularized, be equal to $V_\epsilon(x) = \mathcal{V} \, \varphi_\epsilon(x)$, i.e.

$$V(x) = \mathcal{V} \lim_{\epsilon \to 0} \varphi_\epsilon(x) \equiv \mathcal{V}\delta(x). \tag{2.61}$$

Equation (2.61) defines the so-called *Dirac's delta function* as a limit of φ_ϵ.

When studying Schrödinger equation regularized as done above, it is possible to show, by integrating the differential equation between $-\epsilon$ and ϵ, that in presence of a potential barrier proportional to the Dirac delta function the wave function stays continuous but its derivative has a first kind discontinuity of amplitude

$$\lim_{\epsilon \to 0}(\psi_E'(\epsilon) - \psi_E'(-\epsilon)) = \frac{2m}{\hbar^2}\mathcal{V}\psi_E(0). \tag{2.62}$$

Notice that a potential barrier proportional to the Dirac delta function can be represented equally well by a square barrier of height $\mathcal{V}/L$ and width L, in the limit as $L \to 0$ with $\int_{-\infty}^{\infty} dx V(x) = \mathcal{V}$ kept constant.

2.5.2 The Square Barrier

Let us consider the stationary Schrödinger equation (2.56) with a potential corresponding to the square barrier described above, that is $V(x) = V$ for

$0 < x < L$ and vanishing elsewhere. As in the classical case we can distinguish two different regimes:
a) the case $E > V$, in which classically the flux would be entirely transmitted;
b) the opposite case, $E < V$, in which classically the flux would be entirely reflected.
Let us start with case (a) and distinguish three different regions:
1) The region $x < 0$, in which the general solution is

$$\psi_E(x) = a_+ e^{i\sqrt{2mE}\ x/\hbar} + a_- e^{-i\sqrt{2mE}\ x/\hbar} . \tag{2.63}$$

This wave function corresponds to two opposite fluxes, the first moving rightwards and equal to $|a_+|^2 \sqrt{2E/m}$, the other opposite to the first and equal to $-|a_-|^2 \sqrt{2E/m}$. Since we want to study a quantum process analogous to that described classically, we arbitrarily choose $a_+ = 1$, thus fixing the incident flux to $\sqrt{2E/m}$, hence

$$\psi_E(x) = e^{i\sqrt{2mE}\ x/\hbar} + a\ e^{-i\sqrt{2mE}\ x/\hbar} ; \tag{2.64}$$

a takes into account the possible reflected flux, $|a|^2 \sqrt{2E/m}$. The physically interesting quantity is the fraction of the incident flux which is reflected, which is called the *reflection coefficient* of the barrier and, with our normalization for the incident flux, is $R = |a|^2$.
2) The region $0 < x < L$, where the general solution is

$$\psi_E(x) = b\ e^{i\sqrt{2m(E-V)}\ x/\hbar} + c\ e^{-i\sqrt{2m(E-V)}\ x/\hbar} . \tag{2.65}$$

3) The region $x > L$, where the general solution is given again by (2.63). However, since we want to study reflection and transmission through the barrier, we exclude the possibility of a backward flux, i.e. coming from $x = \infty$, thus assuming that the only particles present in this region are those going rightwards after crossing the barrier. Therefore in this region we write

$$\psi_E(x) = d\ e^{i\sqrt{2mE}\ x/\hbar} . \tag{2.66}$$

The potential has two discontinuities in $x = 0$ and $x = L$, therefore we have the following conditions for the continuity of the wave function and its derivative:

$$1 + a = b + c ,$$

$$1 - a = \sqrt{\frac{E-V}{E}}\,(b - c) ,$$

$$b\ e^{i\sqrt{2m(E-V)}L/\hbar} + c\ e^{-i\sqrt{2m(E-V)}L/\hbar} = d\ e^{i\sqrt{2mE}L/\hbar} ,$$

$$\sqrt{\frac{E-V}{E}}\left[b\ e^{i\sqrt{2m(E-V)}L/\hbar} - c\ e^{-i\sqrt{2m(E-V)}L/\hbar} \right] = d\ e^{i\sqrt{2mE}L/\hbar} \tag{2.67}$$

We have thus a linear system of 4 equations with 4 unknown variables which, for a generic choice of parameters, should univocally identify the solution. However our main interest is the determination of $|a|^2$. Dividing side by side the first two as well as the last two equations, we obtain after simple algebra:

$$\frac{1-a}{1+a} = \sqrt{\frac{E-V}{E}}\frac{\frac{b}{c}-1}{\frac{b}{c}+1},$$

$$\frac{\frac{b}{c}-e^{-2i\sqrt{2m(E-V)}L/\hbar}}{\frac{b}{c}+e^{-2i\sqrt{2m(E-V)}L/\hbar}} = \sqrt{\frac{E}{E-V}}. \tag{2.68}$$

Solving the second equation for b/c and the first for a we obtain:

$$\frac{b}{c} = e^{-2i\sqrt{2m(E-V)}L/\hbar}\frac{\sqrt{\frac{E-V}{E}}+1}{\sqrt{\frac{E-V}{E}}-1},$$

$$a = \frac{1+\sqrt{\frac{E-V}{E}}+\frac{b}{c}\left(1-\sqrt{\frac{E-V}{E}}\right)}{1-\sqrt{\frac{E-V}{E}}+\frac{b}{c}\left(1+\sqrt{\frac{E-V}{E}}\right)} \tag{2.69}$$

and finally, by substitution:

$$a = \frac{\left(1-\frac{E-V}{E}\right)\left(e^{i\sqrt{2m(E-V)}L/\hbar}-e^{-i\sqrt{2m(E-V)}L/\hbar}\right)}{\left(1-\sqrt{\frac{E-V}{E}}\right)^2 e^{i\sqrt{2m(E-V)}L/\hbar}-\left(1+\sqrt{\frac{E-V}{E}}\right)^2 e^{-i\sqrt{2m(E-V)}L/\hbar}},$$

so that

$$a = \frac{V}{E}\frac{\sin\left(\frac{\sqrt{(2m(E-V)}}{\hbar}L\right)}{\frac{2E-V}{E}\sin\left(\frac{\sqrt{(2m(E-V)}}{\hbar}L\right)+2i\sqrt{\frac{E-V}{E}}\cos\left(\frac{\sqrt{(2m(E-V)}}{\hbar}L\right)}, \tag{2.70}$$

which clearly shows that $0 \leq |a| < 1$ and that, for $V > 0$, a vanishes only when $\sqrt{(2m(E-V)}L/\hbar = n\pi$.

This is a clear interference effect showing that reflection by the barrier is a wavelike phenomenon. For those knowing the physics of coaxial cables there should be a clear analogy between our result and the reflection happening at the junction of two cables having mismatching impedances: television set technicians well known that as a possible origin of failure.

The quantum behavior in case (b), i.e. when $E < V$, is more interesting and important for its application to microscopic physics. In this case the wave functions in regions 1 and 3 do not change, while for $0 < x < L$ the general solution is:

$$\psi_E(x) = b\ e^{\sqrt{2m(V-E)}\ x/\hbar} + c\ e^{-\sqrt{2m(V-E)}\ x/\hbar}, \qquad (2.71)$$

so that the continuity conditions become:

$$1 + a = b + c\,,$$

$$1 - a = -i\sqrt{\frac{V-E}{E}}\,(b - c)\,,$$

$$b\ e^{\sqrt{2m(V-E)}L/\hbar} + c\ e^{-\sqrt{2m(V-E)}L/\hbar} = d\ e^{i\sqrt{2mE}L/\hbar}, \qquad (2.72)$$

$$-i\sqrt{\frac{V-E}{E}}\left[b\ e^{\sqrt{2m(V-E)}L/\hbar} - c\ e^{-\sqrt{2m(V-E)}L/\hbar}\right] = d\ e^{i\sqrt{2mE}L/\hbar}\,.$$

Dividing again side by side we have:

$$\frac{\frac{b}{c} - e^{-2\sqrt{2m(V-E)}L/\hbar}}{\frac{b}{c} + e^{-2\sqrt{2m(V-E)}L/\hbar}} = i\sqrt{\frac{E}{V-E}}\,,$$

$$\frac{1-a}{1+a} = i\sqrt{\frac{V-E}{E}}\,\frac{1-\frac{b}{c}}{1+\frac{b}{c}}\,, \qquad (2.73)$$

which can be solved as follows:

$$a = -\frac{1 - \frac{b}{c} + i\sqrt{\frac{E}{V-E}}\left(1+\frac{b}{c}\right)}{1 - \frac{b}{c} - i\sqrt{\frac{E}{V-E}}\left(1+\frac{b}{c}\right)}\,,$$

$$\frac{b}{c} = e^{-2\sqrt{2m(V-E)}L/\hbar}\,\frac{1 + i\sqrt{\frac{E}{V-E}}}{1 - i\sqrt{\frac{E}{V-E}}}\,. \qquad (2.74)$$

We can get the expression for a, hence the reflection coefficient $R \equiv |a|^2$, by replacing b/c in the first equation. The novelty is that R is not equal to one since, as it is clear from (2.74), b/c is a complex number. Therefore a fraction $1 - R \equiv T$ of the incident flux is transmitted through the barrier, in spite of the fact that, classically, the particles do not have enough energy to reach the top of it. That is known as *tunnel effect* and plays a very important role in several branches of modern physics, from radioactivity to electronics.

Instead of giving a complete solution for a, hence for the *transmission coefficient* T, and in order to avoid too complex and unreadable formulae, we will confine the discussion to two extreme cases, which however have a great phenomenological interest. We consider in particular:

a) the case in which $e^{-2\sqrt{2m(V-E)}L/\hbar} \ll 1$, with a generic value for $\frac{E}{V-E}$, i.e. $L \gg \hbar/\sqrt{2m(V-E)}$, which is known as the thick barrier case;

b) the case in which the barrier is thin, corresponding in particular to the limit $L \to 0$ with $VL \equiv \mathcal{V}$ kept constant.

The thick barrier

In this case $|b/c|$ is small, so that it could be neglected in a first approximation, however it is clear from (2.74) that if $b/c = 0$ then $|a| = 1$, so that there is actually no tunnel effect. For this reason we must compute the Taylor expansion in the expression of a as a function of b/c up to the first order:

$$
\begin{aligned}
a &= -\frac{1 + i\sqrt{\frac{E}{V-E}}\ 1 - \frac{b}{c}\frac{1 - i\sqrt{E/(V-E)}}{1 + i\sqrt{E/(V-E)}}}{1 - i\sqrt{\frac{E}{V-E}}\ 1 - \frac{b}{c}\frac{1 + i\sqrt{E/(V-E)}}{1 - i\sqrt{E/(V-E)}}} \\[2ex]
&\sim -\frac{1 + i\sqrt{\frac{E}{V-E}}}{1 - i\sqrt{\frac{E}{V-E}}}\left[1 - \frac{b}{c}\left(\frac{1 - i\sqrt{\frac{E}{V-E}}}{1 + i\sqrt{\frac{E}{V-E}}} - \frac{1 + i\sqrt{\frac{E}{V-E}}}{1 - i\sqrt{\frac{E}{V-E}}}\right)\right] \\[2ex]
&= -\frac{1 + i\sqrt{\frac{E}{V-E}}}{1 - i\sqrt{\frac{E}{V-E}}}\left[1 + 4i\frac{b}{c}\frac{\sqrt{E(V-E)}}{V}\right] \\[2ex]
&= -\frac{1 + i\sqrt{\frac{E}{V-E}}}{1 - i\sqrt{\frac{E}{V-E}}}\left[1 + 4i\frac{\sqrt{E(V-E)}}{V}e^{-2\sqrt{2m(V-E)}L/\hbar}\frac{1 + i\sqrt{\frac{E}{V-E}}}{1 - i\sqrt{\frac{E}{V-E}}}\right].
\end{aligned}
\tag{2.75}
$$

In the last line we have replaced b/c by the corresponding expression in (2.74). Neglecting terms of the order of $e^{-4\sqrt{2m(V-E)}L/\hbar}$ or smaller we obtain

$$
|a|^2 = R = 1 - 16\frac{E(V-E)}{V^2}e^{-2\sqrt{2m(V-E)}L/\hbar}.
\tag{2.76}
$$

Therefore the transmission coefficient, which measures the probability for a particle hitting the barrier to cross it, is given by:

$$
T \equiv 1 - R = 16\frac{E(V-E)}{V^2}e^{-2\sqrt{2m(V-E)}L/\hbar}.
\tag{2.77}
$$

Notice that the result seems to vanish for $V = E$, but this is not true since in this case the terms neglected in our approximation come into play.

This formula was first applied in nuclear physics, and more precisely to study α emission, a phenomenon in which a heavy nucleus breaks up into a lighter nucleus plus a particle carrying twice the charge of the proton and roughly four times its mass, which is known as α particle. The decay can be simply described in terms of particles of mass $0.7\ 10^{-26}$ Kg and energy $E \simeq 4 - 8$ MeV $\simeq 10^{-12}$ J, hitting barriers of width roughly equal to $3\ 10^{-14}$ m; the difference $V - E$ is of the order of 10 MeV $= 2\ 10^7$ eV $\simeq 3.2\ 10^{-12}$ J.

In these conditions we have $2\sqrt{2m(V-E)}L/\hbar \simeq 84$ and therefore $T \sim e^{-2\sqrt{2m(V-E)}L/\hbar} \sim 10^{-36}$. Given the order of magnitude of the energy E and of the mass of the particle, we infer that it moves with velocities of the order of

10^7 m/s: since the radius R_0 of heavy nuclei is roughly 10^{-14} m, the frequency of collisions against the barrier is $\nu_u \sim 10^{21}$ Hz. That indicates that, on average, the time needed for the α particle to escape the nucleus is of the order of $1/(\nu_u T)$, i.e. about 10^{15} s, equal to 10^8 years. However, if the width of the barrier is only 4 times smaller, the decay time goes down to about 100 years. That shows a great sensitivity of the result to the parameters and justifies the fact that we have neglected the pre-factor in front of the exponential in (2.77). On the other hand that also shows that, for a serious comparison with the actual mean lives of nuclei, an accurate analysis of parameters is needed, but it is also necessary to take into account the fact that we are not dealing with a true square barrier, since the repulsion between the nucleus and the α particle is determined by Coulomb forces, i.e. $V(x) = 2Ze^2/(4\pi\epsilon_0 x)$ for x greater than a given threshold, see Fig. 2.3.

As a consequence, the order of magnitude of the transmission coefficient given in (2.77), i.e.

$$T \simeq e^{-2\sqrt{2m(V-E)}L/\hbar} \tag{2.78}$$

must be replaced by[1]

$$T \simeq \exp\left(-2\int_{R_0}^{R_1} dx \frac{\sqrt{2m(V(x)-E)}}{\hbar}\right) \equiv e^{-G}, \tag{2.79}$$

where R_0 is the already mentioned nuclear radius and $R_1 = 2Ze^2/(4\pi E\epsilon_0)$ is the solution of the equation $V(R_1) = E$. We have then

$$G = 2\frac{\sqrt{2m}}{\hbar}\int_{R_0}^{R_1} dx\sqrt{\frac{2Ze^2}{4\pi\epsilon_0 x} - E} = 2\frac{\sqrt{2mE}}{\hbar}\int_{R_0}^{R_1} dx\sqrt{\frac{R_1}{x} - 1}$$

$$= 2\frac{\sqrt{2mE}R_1}{\hbar}\int_{\frac{R_0}{R_1}}^{1} dy\sqrt{\frac{1}{y} - 1} = 2\sqrt{\frac{2m}{E}}\frac{Ze^2}{\pi\epsilon_0\hbar}\int_{\sqrt{\frac{R_0}{R_1}}}^{1} dz\sqrt{1 - z^2}$$

$$= \sqrt{\frac{2m}{E}}\frac{Ze^2}{\pi\epsilon_0\hbar}\left[\text{acos}\sqrt{\frac{R_0}{R_1}} - \sqrt{\frac{R_0}{R_1} - \left(\frac{R_0}{R_1}\right)^2}\right]. \tag{2.80}$$

In the approximation $R_0/R_1 << 1$ we have

$$G \simeq \frac{2\pi Ze^2}{\epsilon_0 hv}, \tag{2.81}$$

where v is the velocity of the alpha particle. Hence, if we assume like above that the collision frequency be $\nu_u \sim 10^{21}$ Hz, the mean life is

---

[1] One can think of a thick but not square barrier as a series of thick square barriers of different heights.

$$\tau = 10^{-21} \exp\left(\frac{2\pi Z e^2}{\epsilon_0 h v}\right). \tag{2.82}$$

If we instead make use of the last expression in (2.80), with $R_0 = 1.1 \; 10^{-14}$ m, we infer for $\ln \tau$ the behavior shown in Fig. 2.4, where the crosses indicates experimental values for the mean lives of various isotopes: ^{232}Th, ^{238}U, ^{230}Th, ^{241}Am, ^{230}U, ^{210}Rn, ^{220}Rn, ^{222}Ac, ^{215}Po, ^{218}Th. Taking into account that the figure covers 23 orders of magnitude, the agreement is surely remarkable. Indeed Gamow's first presentation of these results in 1928 made a great impression.

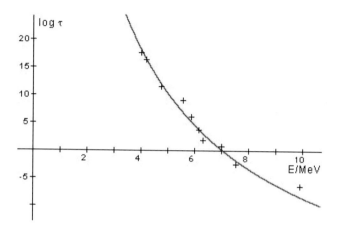

Fig. 2.4. The mean lives of a sample of α-emitting isotopes plotted against the corresponding α-energies. The solid line shows the values foreseen by Gamow's theory

The thin barrier

In the case of a thin barrier we can neglect E with respect to V, so that $\sqrt{E/V - E} \simeq \sqrt{E/V}$ and $e^{-\sqrt{2m(V-E)}L/\hbar}$ can be replaced by $1 - \sqrt{2mV}L/\hbar$. We also remind that $\sqrt{2mV}L/\hbar$ is infinitesimal so that $e^{i\sqrt{2mEL}/\hbar}$ can be put equal to 1. Therefore equation (2.73) becomes

$$1 + a = b + c,$$

$$1 - a = -i\sqrt{\frac{V}{E}}(b - c),$$

$$b + c + \frac{\sqrt{2mV}}{\hbar}L(b - c) = d,$$

$$b - c + \frac{\sqrt{2mV}}{\hbar}L(b + c) = i\sqrt{\frac{E}{V}}d, \tag{2.83}$$

and substituting $b \pm c$ we obtain:

$$1 + a + i\sqrt{\frac{E}{V}} \frac{\sqrt{2mV}}{\hbar} L(1-a) = d \simeq 1 + a \,,$$

$$i\sqrt{\frac{E}{V}}(1-a) + \frac{\sqrt{2mV}}{\hbar} L(1+a) = i\sqrt{\frac{E}{V}} d \,, \qquad (2.84)$$

in its simplest form. Taking further into account our approximation, the system can be rewritten as

$$1 + a = d \,,$$

$$1 - a = \left(i\sqrt{\frac{2m}{E}} \frac{VL}{\hbar} + 1\right) d \equiv \left(1 + i\sqrt{\frac{2m}{E}} \frac{V}{\hbar}\right) d \,. \qquad (2.85)$$

Finally we find, by eliminating a, that

$$d = \frac{1}{1 + i\sqrt{\frac{m}{2E}} \frac{V}{\hbar}} \,, \qquad (2.86)$$

hence

$$T = \frac{1}{1 + \frac{m}{2E} \frac{V^2}{\hbar^2}} \qquad (2.87)$$

and

$$R = \frac{1}{1 + \frac{2E}{m} \frac{\hbar^2}{V^2}} \,. \qquad (2.88)$$

Notice that the system (2.85) confirms what predicted about the continuity conditions for the wave function in presence of a potential energy equal to $V\delta(x)$, i.e. that the wave function is continuous $(1 + a = d)$ while its derivative has a discontinuity $(i\sqrt{2mE}(1 - a - d)/\hbar)$ equal to $2mV/\hbar^2$ times the value of the wave function (d in our case).

2.6 Quantum Wells and Energy Levels

Having explored the tunnel effect in some details, let us now discuss the solution of Schrödinger's equation in the case of binding potentials. For bound states, i.e. for solutions with wave functions localized in the neighborhood of the well, we expect computations to lead to energy quantization, i.e. to the presence of discrete *energy levels*. Let us start our discussion from the case of a square well

$$V(x) = -V \quad \text{for} \quad |x| < \frac{L}{2} \,, \qquad V(x) = 0 \quad \text{for} \quad |x| > \frac{L}{2} \,. \qquad (2.89)$$

Notice that the origin of the co-ordinate has been chosen in order to emphasize the *symmetry* of the system, corresponding in this case to the invariance of Schrödinger equation under axis reflection $x \to -x$. In general the symmetry of the potential allows us to find new

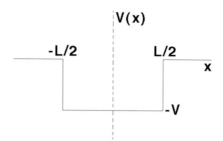

solutions of the equation starting from known solutions, or to simplify the search for solutions by a-priori fixing some of their features. In this case it can be noticed that if $\psi_E(x)$ is a solution, $\psi_E(-x)$ is a solution too, so that, by linearity of the differential equation, any linear combination (with complex coefficients) of the two wave functions is a good solution corresponding to the same value of the energy E, in particular the combinations $\psi_E(x) \pm \psi_E(-x)$, which are even/odd under reflection of the x axis. Naturally one of the two solutions may well vanish, but it is clear that all possible solutions can be described in terms of (i.e. they can be written as linear combinations of) functions which are either even or odd under x-reflection.

To better clarify the point, let us notice that, since the Schrödinger equation is linear, the set of all its possible solutions with the same energy constitutes what is usually called a *linear space*, which is completely fixed once we know one particular basis for it. What we have learned is that in the present case even/odd functions are a good basis, so that the search for solutions can be solely limited to them. This is probably the simplest example of the application of a *symmetry principle* asserting that, if the Schrödinger equation is invariant under a coordinate transformation, it is always possible to choose its solutions so that the transformation does not change them but for a constant phase factor, which in the present case is ± 1.

We will consider in the following only bound solutions which, assuming that the potential energy vanishes as $|x| \to \infty$, correspond to a negative total energy E and are therefore the analogous of bound states in classical mechanics. Solutions with positive energy instead present reflection and transmission phenomena as in the case of barriers. We notice that, in the case of bound states, the collective interpretation of the wave function does not apply, since these are states involving a single particle: that is in strict relation with the fact that the bound state solutions vanish rapidly enough as $|x| \to \infty$ so that the probability distribution in (2.44) can be properly normalized.

Let us start by considering even solutions: it is clear that we can limit our study to the positive x axis, with the additional constraint of a vanishing first derivative in the origin, as due for an even function (whose derivative is odd). We can divide the positive x axis into two regions where the potential is constant:

a) That corresponding to $x < L/2$, where the general solution is:

$$\psi_E(x) = a_+ e^{i\sqrt{2m(E+V)}\ x/\hbar} + a_- e^{-i\sqrt{2m(E+V)}\ x/\hbar} ,$$

which is even for $a_+ = a_-$, so that

$$\psi_E(x) = a \cos\left(\frac{\sqrt{2m(E+V)}}{\hbar} x\right) . \tag{2.90}$$

b) That corresponding to $x > L/2$, where the general solution is:

$$\psi_E(x) = b_+ e^{\sqrt{2m|E|}\ x/\hbar} + b_- e^{-\sqrt{2m|E|}\ x/\hbar} .$$

The condition that $|\psi|^2$ be an integrable function constrains $b_+ = 0$, otherwise the probability density would unphysically diverge as $|x| \to \infty$; therefore we can write

$$\psi_E(x) = b\ e^{-\sqrt{2m|E|}\ x/\hbar} . \tag{2.91}$$

Notice that we have implicitly excluded the possibility $E < -V$, the reason being that in this case (2.90) would be replaced by

$$\psi_E(x) = a \cosh\left(\frac{\sqrt{2m|E+V|}}{\hbar} x\right)$$

which for $x > 0$ has a positive logarithmic derivative $(\partial_x \psi_E(x)/\psi_E(x))$ which cannot continuously match the negative logarithmic derivative of the solution in the second region given in (2.91). Therefore quantum theory is in agreement with classical mechanics about the impossibility of having states with total energy less than the minimum of the potential energy.

The solutions of the Schrödinger equation on the whole axis can be found by solving the system:

$$a \cos \frac{\sqrt{2m(E+V)}L}{2\hbar} = b\ e^{-\sqrt{2m|E|}L/(2\hbar)} ,$$

$$\frac{\sqrt{2m(E+V)}}{\hbar} a \sin \frac{\sqrt{2m(E+V)}L}{2\hbar} = \frac{\sqrt{2m|E|}}{\hbar} b\ e^{-\sqrt{2m|E|}L/(2\hbar)} \tag{2.92}$$

dividing previous equations side by side we obtain the continuity condition for the logarithmic derivative:

$$\tan \frac{\sqrt{2m(E+V)}L}{2\hbar} = \sqrt{\frac{|E|}{E+V}} . \tag{2.93}$$

In order to discuss last equation let us introduce the variable

$$x \equiv \frac{\sqrt{2m(E+V)}L}{2\hbar} \tag{2.94}$$

and the parameter

$$y \equiv \sqrt{2mV}L/2\hbar \ , \qquad (2.95)$$

and let us plot together the behavior of the two functions $\tan x$ and $\sqrt{(y^2 - x^2)/x^2} = \sqrt{|E|/(E+V)}$. In the figure we show the case $y^2 = 20$. From a qualitative point of view the figure shows that energy levels, corresponding to the intersection points of the two functions, are quantized, thus confirming also for the case of potential wells the discrete energy spectrum predicted by Bohr's theory. In particular the plot shows two intersections, the first for $x = x_1 < \pi/2$, the second for $\pi < x = x_2 < 3\pi/2$.

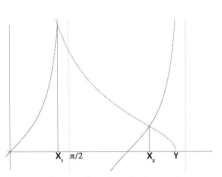

Notice that quantization of energy derives from the physical requirement of having a bound state solution which does not diverge but instead vanishes outside the well: for this reason the external solution is parametrized in terms of only one parameter. The reduced number of available parameters allows for non-trivial solution of the homogeneous linear system (2.92) only if the energy quantization condition (2.93) is satisfied.

The number of possible solutions increases as y grows and since $y > 0$ it is anyway greater than zero. Therefore the square potential well in one dimension has always at least one bound state corresponding to an even wave function. It can be proved that the same is true for every symmetric well in one dimension (i.e. such that $V(-x) = V(x)$). On the contrary an extension of this analysis shows that in the three dimensional case the existence of at least one bound state is not guaranteed any more.

Let us now consider the case of odd solutions: we must choose a wave function which vanishes in the origin, so that the cosine must be replaced by a sine in (2.90). Going along the same lines leading to (2.93) we arrive to the equation

$$\cot \frac{\sqrt{2m(E+V)}L}{2\hbar} = -\sqrt{\frac{|E|}{E+V}} \ . \qquad (2.96)$$

Using the same variables x and y as above, we have the corresponding figure on the side, which shows that intersections are present only if $y > \pi/2$, i.e. if $V > \pi^2\hbar^2/(2mL^2)$ (which by the way is also the condition for the existence of at least one bound state in three dimensions). Notice that the energy levels found

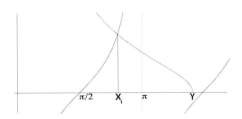

in the odd case are different from those found in the even case. In particular any possible negative energy level can be put in correspondence with only one wave function (identified by neglecting a possible irrelevant constant phase factor): this implies that, in the present case, dealing with solutions having a definite transformation property under the symmetry of the problem (i.e. even or odd) is not a matter of choice, as it is in the general case, but a necessity, since those are the only possible solutions. Indeed a different kind of solution could only be constructed in presence of two solutions, one even and the other odd, corresponding to the same energy level.

The number of independent solutions corresponding to a given energy level is usually called the *degeneracy* of the level. We have therefore demonstrated that, for the potential square well in one dimension, the discrete energy levels have always degeneracy equal to one or, stated otherwise, that they are *non-degenerate*. This is in fact a general property of bound states in one dimension, which can be demonstrated for any kind of potential well.

It is interesting to apply our analysis to the case of an infinitely deep well. Obviously, if we want to avoid dealing with divergent negative energies as we deepen the well, it is convenient to shift the zero of the energy so that the potential energy vanish inside the well and be V outside. That is equivalent to replacing in previous formulae $E + V$ by E and $|E|$ by $V - E$; moreover, bound states will now correspond to energies $E < V$. Taking the limit $V \to \infty$ in the quantization conditions given in (2.93) and (2.96), we obtain respectively $\tan \sqrt{2mEL}/(2\hbar) = +\infty$ and $-\cot \sqrt{2mEL}/(2\hbar) = +\infty$, so that $\sqrt{2mEL}/(2\hbar) = (2n-1)\pi/2$ and $\sqrt{2mEL}/(2\hbar) = n\pi$ with $n = 1, 2, \ldots$ Finally, combining odd and even states, we have

$$\frac{\sqrt{2mE}}{\hbar} L = n\pi \quad : \quad n = 1, 2, \ldots$$

and the following energy levels

$$E_n = \frac{n^2\pi^2\hbar^2}{2mL^2} . \tag{2.97}$$

The corresponding wave functions vanish outside the well while in the region $|x| < L/2$ the even functions are $\sqrt{2/L}\cos(2n-1)\pi x/L$ and the odd ones are $\sqrt{2/L}\sin 2n\pi x/L$, with the coefficients fixed in order to satisfy (2.44). It is also possible to describe all wave functions by a unique formula:

$$\psi_{E_n}(x) = \sqrt{\frac{2}{L}} \sin \frac{n\pi(x + \frac{L}{2})}{L} \quad \text{for} \quad |x| < \frac{L}{2},$$

$$\psi_{E_n}(x) = 0 \quad \text{for} \quad |x| > \frac{L}{2} . \tag{2.98}$$

While all wave functions are contin-
uos in $|x| = L/2$, their derivatives are
not, as in the case of the potential
barrier proportional to the Dirac delta
function. The generic solution ψ_{E_n} has
the behaviour showed in the figure,
where the analogy with the electric
component of an electromagnetic wave
reflected between two mirrors clearly
appears. Therefore the infinitely deep
well can be identified as the region between two reflecting walls.

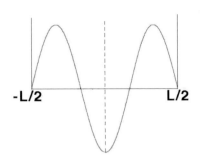

If the wave amplitude vanishes over the mirrors, the distance between
them must necessarily be an integer multiple of half the wavelength; this is
the typical tuning condition for a musical instrument and implies wavelength
and energy quantization. The exact result agrees with that of Problem 2.4.

Going back to the analogy with electromagnetic waves, the present situation corresponds to a one-dimensional *resonant cavity*. In the cavity the field
can only oscillate according to the permitted wavelengths, which are $\lambda_n = 2L/n$ for $n = 1, 2, \ldots$ corresponding to the frequencies $\nu_n = c/\lambda_n = nc/(2L)$,
which are all multiple of the *fundamental* frequency of the cavity.

Our results regarding the infinitely deep well can be easily generalized to
three dimensions. To that purpose, let us introduce a cubic box of side L with
reflecting walls. The conditions that the wave function vanish over the walls
are equivalent, inside the box, to:

$$\psi_{n_x,n_y,n_z} = \sqrt{\frac{8}{L^3}} \sin \frac{n_x \pi (x + \frac{L}{2})}{L} \sin \frac{n_y \pi (y + \frac{L}{2})}{L} \sin \frac{n_z \pi (z + \frac{L}{2})}{L} . \qquad (2.99)$$

where we have assumed the origin of the coordinates to be placed in the center
of the box. The corresponding energy coincides with the kinetic energy inside
the box and can be obtained by writing the Schrödinger equation in three
dimensions:

$$-\frac{\hbar^2}{2m} \left(\partial_x^2 + \partial_y^2 + \partial_z^2 \right) \psi_{n_x,n_y,n_z} = E_{n_x,n_y,n_z} \psi_{n_x,n_y,n_z} , \qquad (2.100)$$

leading to

$$E_{n_x,n_y,n_z} = \frac{\pi^2 \hbar^2}{2mL^2} \left[n_x^2 + n_y^2 + n_z^2 \right] . \qquad (2.101)$$

This result will be useful for studying the properties of a gas of non-interacting
particles (perfect gas) contained in a box with reflecting walls. Following the
same analogy as above one can study in a similar way the oscillations of an
electromagnetic field in a three-dimensional cavity, with proper frequencies
given by $\nu_{n_x,n_y,n_z} = (c/2L)\sqrt{n_x^2 + n_y^2 + n_z^2}$.

2.7 The Harmonic Oscillator

The one-dimensional harmonic oscillator can be identified with the mechanical system formed by a particle of mass m bound to a fixed point (taken as the origin of the coordinate) by an ideal spring of elastic constant k and vanishing length at rest. This is equivalent to a potential energy $V(x) = kx^2/2$. In classical mechanics the corresponding equation of motion is

$$m\ddot{x} + kx = 0\,,$$

whose general solution is

$$x(t) = X\cos(\omega t + \phi)\,,$$

where $\omega = \sqrt{k/m} = 2\pi\nu$ and ν is the proper frequency of the oscillator.

At the quantum level we must solve the following stationary Schrödinger equation:

$$-\frac{\hbar^2}{2m}\partial_x^2\psi_E(x) + \frac{k}{2}x^2\psi_E(x) = E\psi_E(x)\,. \tag{2.102}$$

In order to solve this equation we can use the identity

$$\left(\sqrt{\frac{k}{2}}x - \frac{\hbar}{\sqrt{2m}}\partial_x\right)\left(\sqrt{\frac{k}{2}}x + \frac{\hbar}{\sqrt{2m}}\partial_x\right)f(x)$$

$$\equiv \frac{k}{2}x^2 f(x) + \frac{\hbar\omega}{2}x\partial_x f(x) - \frac{\hbar\omega}{2}\partial_x(xf(x)) - \frac{\hbar^2}{2m}\partial_x^2 f(x)$$

$$= -\frac{\hbar^2}{2m}\partial_x^2 f(x) + \frac{k}{2}x^2 f(x) - \frac{\hbar\omega}{2}f(x)$$

$$\equiv \left(-\frac{\hbar^2}{2m}\partial_x^2 + \frac{k}{2}x^2 - \frac{\hbar\omega}{2}\right)f(x)\,, \tag{2.103}$$

which is true for any function f which is derivable at least two times.

It is important to notice the *operator* notation used in last equation, where we have introduce some specific symbols, $(\sqrt{k/2}x \pm (\hbar/\sqrt{2m})\partial_x)$ or $(-(\hbar^2/2m)\partial_x^2 + (k/2)x^2 - \hbar\omega/2)$, to indicate operations in which derivation and multiplication by some variable are combined together. These are usually called *operators*, meaning that they give a correspondence law between functions belonging to some given class (for instance those which can be derived n times) and other functions belonging, in general, to a different class.

In this way, leaving aside the specific function f, equation (2.103) can be rewritten as an operator relation

$$\left(\sqrt{\frac{k}{2}}x - \frac{\hbar}{\sqrt{2m}}\partial_x\right)\left(\sqrt{\frac{k}{2}}x + \frac{\hbar}{\sqrt{2m}}\partial_x\right) = \left(-\frac{\hbar^2}{2m}\partial_x^2 + \frac{k}{2}x^2 - \frac{\hbar\omega}{2}\right) \tag{2.104}$$

and equations of similar nature can be introduced, like for instance:

$$\left(\sqrt{\frac{k}{2}}x + \frac{\hbar}{\sqrt{2m}}\partial_x\right)\left(\sqrt{\frac{k}{2}}x - \frac{\hbar}{\sqrt{2m}}\partial_x\right)$$

$$- \left(\sqrt{\frac{k}{2}}x - \frac{\hbar}{\sqrt{2m}}\partial_x\right)\left(\sqrt{\frac{k}{2}}x + \frac{\hbar}{\sqrt{2m}}\partial_x\right) = \hbar\omega. \qquad (2.105)$$

In order to shorten formulae, it is useful to introduce the two symbols:

$$X_\pm \equiv \left(\sqrt{\frac{k}{2}}x \pm \frac{\hbar}{\sqrt{2m}}\partial_x\right). \qquad (2.106)$$

That allows us to rewrite (2.105) in the simpler form:

$$X_+X_- - X_-X_+ = \hbar\omega. \qquad (2.107)$$

If, extending the operator formalism, we define

$$H \equiv -\frac{\hbar^2}{2m}\partial_x^2 + \frac{k}{2}x^2, \qquad (2.108)$$

we can rewrite (2.104) as:

$$X_-X_+ = H - \frac{\hbar\omega}{2}, \qquad (2.109)$$

then obtaining from (2.107):

$$X_+X_- = H + \frac{\hbar\omega}{2}. \qquad (2.110)$$

The Schrödinger equation can be finally written as:

$$H\psi_E(x) = E\psi_E(x). \qquad (2.111)$$

The operator formalism permits to get quite rapidly a series of results:
   a) The wave function which is solution of the equation

$$X_+\psi_0(x) = \sqrt{\frac{k}{2}}x\psi_0(x) + \frac{\hbar}{\sqrt{2m}}\partial_x\psi_0(x) = 0, \qquad (2.112)$$

is also a solution of (2.111) with $E = \hbar\omega/2$. In order to compute it we can rewrite (2.112) as:

$$\frac{\partial_x\psi_0(x)}{\psi_0(x)} = -\frac{\sqrt{km}}{\hbar}x,$$

hence, integrating both members:

$$\ln\psi_0(x) = c - \frac{\sqrt{km}}{2\hbar}x^2,$$

from which it follows that

$$\psi_0(x) = e^c \, e^{-\sqrt{km} \, x^2/(2\hbar)} \, ,$$

where the constant c can be fixed by the normalization condition given in (2.44), leading finally to

$$\psi_0(x) = \left(\frac{km}{\pi^2\hbar^2} \right)^{\frac{1}{8}} e^{-\sqrt{km} \, x^2/\hbar} \, . \qquad (2.113)$$

We would like to remind the need for restricting the analysis to the so-called *square integrable* functions, which can be normalized according to (2.44). This is understood in the following.

b) What we have found is the lowest energy solution, usually called the *fundamental state* of the system, as can be proved by observing that, for every normalized solution $\psi_E(x)$, the following relations hold:

$$\int_{-\infty}^{\infty} dx \, \psi_E(x)^* \left(\sqrt{\frac{k}{2}} x - \frac{\hbar}{\sqrt{2m}} \partial_x \right) \left(\sqrt{\frac{k}{2}} x + \frac{\hbar}{\sqrt{2m}} \partial_x \right) \psi_E(x)$$

$$= \int_{-\infty}^{\infty} dx |X_+\psi_E(x)|^2 = \int_{-\infty}^{\infty} dx \, \psi_E(x)^* \left(E - \frac{\hbar\omega}{2} \right) \psi_E(x)$$

$$= E - \frac{\hbar\omega}{2} \geq 0 \, . \qquad (2.114)$$

where the derivative in X_- has been integrated by parts. Last inequality follows from the fact that the integral of the squared modulus of any function cannot be negative. Moreover it must be noticed that if the integral vanishes, i.e. if $E = \hbar\omega/2$, then necessarily $X_+\psi_E = 0$, so that ψ_E is proportional to ψ_0. That proves that the fundamental state is unique.

c) If ψ_E satisfies (2.102) then $X_\pm\psi_E$ satisfies the same equation with E replaced by $E \mp \hbar\omega$, i.e. we have

$$HX_\pm\psi_E = (E \mp \hbar\omega)X_\pm\psi_E \, . \qquad (2.115)$$

Notice that $X_+\psi_E$ vanishes if and only if $\psi_E = \psi_0$ while $X_-\psi_E$ never vanishes: one can prove this by verifying that if $X_-\psi_E = 0$ then ψ_E behaves as ψ_0 but with the sign + in the exponent, hence it is not square integrable. In order to prove equation (2.115), from (2.109) and (2.110) we infer, for instance:

$$X_+X_-X_+\psi_E = X_+ \left(H - \frac{\hbar\omega}{2} \right) \psi_E(x) = \left(H + \frac{\hbar\omega}{2} \right) X_+\psi_E(x)$$

$$= X_+ \left(E - \frac{\hbar\omega}{2} \right) \psi_E(x) = \left(E - \frac{\hbar\omega}{2} \right) X_+\psi_E(x) \quad (2.116)$$

from which (2.115) follows in the + case.

Last computations show that operators combine in a fashion which resembles usual multiplication, however their product is strictly dependent on the order in which they appear. We say that the product is *non-commutative*; that is also evident from (2.107).

Exchanging X_- and X_+ in previous equations we have:

$$X_- X_+ X_- \psi_E = X_- \left(H + \frac{\hbar\omega}{2} \right) \psi_E(x) = \left(H - \frac{\hbar\omega}{2} \right) X_- \psi_E(x)$$

$$= X_- \left(E + \frac{\hbar\omega}{2} \right) \psi_E(x) = \left(E + \frac{\hbar\omega}{2} \right) X_- \psi_E(x) \quad (2.117)$$

which completes the proof of (2.115).

d) Finally, combining points (a–c), we can show that the only possible energy levels are:

$$E_n = \left(n + \frac{1}{2} \right) \hbar\omega . \quad (2.118)$$

In order to prove that, let us suppose instead that (2.102) admits the level $E = (m + 1/2)\hbar\omega + \delta$, where $0 < \delta < \hbar\omega$, and then repeatedly apply X_+ to ψ_E up to $m+1$ times. If $X_+^k \psi_E = 0$ with $k \le m+1$ and $X_+^{k-1} \psi_E \ne 0$, then we would have $X_+(X_+^{k-1} \psi_E) = 0$ which, as we have already seen, is equivalent to $X_+^{k-1} \psi_E \sim \psi_0$, hence to $H X_+^{k-1} \psi_E = \hbar\omega/2 \psi_E$. However equation (2.115) implies $H X_+^{k-1} \psi_E = (E - (k-1)\hbar\omega) \psi_E$, hence $E = (k - 1/2)\hbar\omega$, which is in contrast with the starting hypothesis ($\delta \ne 0$). On the other hand $X_+^k \psi_E \ne 0$ even for $k = m + 1$ would imply the presence of a solution with energy less that $\hbar\omega/2$, in contrast with (2.114). We have instead no contradiction if $\delta = 0$ and $k = m + 1$.

We have therefore shown that the spectrum of the harmonic oscillator consists of the energy levels $E_n = (n + 1/2)\hbar\omega$. We also know from (2.115) that $\sim X_-^n \psi_0$ is a possible solution with $E = E_n$: we will now show that this is actually the only possible solution.

e) Any wave function corresponding to the n-th energy level is necessarily proportional to $X_-^n \psi_0$:

$$\psi_{E_n} \sim X_-^n \psi_0 . \quad (2.119)$$

We already know that this is true for $n = 0$ (fundamental state). Now let us suppose the same to be true for $n = k$ and we will prove it for $n = k+1$, thus concluding our argument by induction. Let $\psi_{E_{k+1}}$ be a solution corresponding to E_{k+1}, then, by (2.115) and by the uniqueness of ψ_{E_k}, we have

$$X_+ \psi_{E_{k+1}} = a \psi_{E_k} \quad (2.120)$$

for some constant $a \ne 0$, with $\psi_{E_k} \propto X_-^k \psi_0$. By applying X_- to both sides of last equation we obtain

$$X_- X_+ \psi_{E_{k+1}} = \left(H - \frac{\hbar\omega}{2} \right) \psi_{E_{k+1}} = (k+1)\,\hbar\omega\,\psi_{E_{k+1}}$$

$$= X_- a \psi_{E_k} \propto X_- X_-^k \psi_0 = X_-^{k+1} \psi_0 , \quad (2.121)$$

which proves that also $\psi_{E_{k+1}}$ is proportional to $X_-^{k+1}\psi_0$.

That concludes our analysis of the one-dimensional harmonic oscillator, which, based on an algebraic approach, has led us to finding both the possible energy levels, given in (2.118), and the corresponding wave functions, described by (2.113) and (2.119). In particular, confirming a general property of bound states in one dimension, we have found that the energy levels are non-degenerate. The operators X_+ and X_- permit to transform a given solution into a different one, in particular by rising (X_-) or lowering (X_+) the energy level by one quantum $\hbar\omega$.

Also in this case, as for the square well, solutions have definite transformation properties under axis reflections $x \to -x$ which follow from the symmetry of the potential, $V(-x) = V(x)$. In particular they are divided in even and odd functions according to the value of n, $\psi_{E_n}(-x) = (-1)^n \psi_{E_n}(x)$, as can be proved by noticing that ψ_0 is an even function and that the operator X_- transforms an even (odd) function into an odd (even) one.

Moreover we notice that, according to (2.119), (2.113) and to the expression for X_- given in (2.106), all wave functions are real. This is also a general property of bound states in one dimension, which can be easily proved and has a simple interpretation. Indeed, suppose ψ_E be the solution of the stationary Schrödinger equation (2.56) corresponding to a discrete energy level E; since obviously both E and the potential energy $V(x)$ are real, it follows, by taking the complex conjugate of both sides of (2.56), that also ψ_E^* is a good solution corresponding to the same energy. However the non-degeneracy of bound states in one dimension implies that ψ_E must be unique. The only possibility is $\psi_E^* \propto \psi_E$, hence $\psi_E^* = e^{i\phi}\psi_E$, so that, leaving aside an irrelevant overall phase factor, ψ_E is a real function.

On the other hand, recalling the definition of the probability current density J given in (2.43), it can be easily proved that the wave function is real if and only if the current density vanishes everywhere. Since we are considering a stationary problem, the probability density is constant in time by definition and the conservation equation (2.27) implies, in one dimension, that the current density J is a constant in space (the same is not true in more than one dimension, where that translates in $\boldsymbol{J}$ being a vector field with vanishing divergence). On the other hand, for a bound state J must surely vanish as $|x| \to \infty$, hence it must vanish everywhere, implying a real wave function: in a one-dimensional bound state there is no current flow at all.

Our results admit various generalization of great physical interest. First of all let us consider their extension to the isotropic three-dimensional harmonic oscillator corresponding to the following Schrödinger equation:

$$-\frac{\hbar^2}{2m}\left(\partial_x^2 + \partial_y^2 + \partial_z^2\right)\psi_E + \frac{k}{2}\left(x^2 + y^2 + z^2\right)\psi_E = E\,\psi_E\,, \quad (2.122)$$

where $\psi_E = \psi_E(x, y, z)$ is the three-dimensional wave function. This equation can be seen as the sum term by term of the equations for three one-dimensional oscillators corresponding to the variables x, y and z. Therefore we conclude that the quantized energy levels are in this case:

$$E_{n_x,n_y,n_z} = \hbar\omega \left(n_x + n_y + n_z + \frac{3}{2} \right) , \qquad (2.123)$$

and that the corresponding wave functions are

$$\psi_{n_x,n_y,n_z}(x,y,z) = \psi_{n_x}(x)\psi_{n_y}(y)\psi_{n_z}(z) . \qquad (2.124)$$

This is the typical example of a *separable* Schrödinger equation. Notice that, according to (2.123) and (2.124), in three dimensions several degenerate solutions can be found for a given energy level, corresponding to all possible integers n_x, n_y, n_z such that $n_x + n_y + n_z = $ const.

A further generalization is that regarding small oscillations around equilibrium for a system with N degrees of freedom, whose energy can be separated into the sum of the contributions from N one-dimensional oscillators having, in general, different proper frequencies (ν_i , $i = 1, ..., N$). In this case the quantization formula reads

$$E_{(n_1,...n_N)} = \sum_{i=1}^{N} \hbar\omega_i \left(n_i + \frac{1}{2} \right) , \qquad (2.125)$$

and the corresponding wave function can be written as the product of the wave functions associated with every single oscillator.

Let us now take a short detour by recalling the analysis of the electromagnetic field resonating in one dimension. It can be shown that, from a dynamical point of view, the electromagnetic field can be described as an ensemble of harmonic oscillators, i.e. mechanical systems with definite frequencies which have been discussed in last Section. Applying the result of this Section we confirm Einstein's assumption that the electromagnetic field can only exchange quanta of energy equal to $\hbar\omega = h\nu$. That justifies the concept of a photon as a particle carrying an energy equal to $h\nu$. At the quantum level the possible states of an electromagnetic field oscillating in a cavity can thus be seen as those of a system of photons, corresponding in number to the total quanta of energy present in the cavity, which bounce elastically between the walls.

2.8 Periodic Potentials and Band spectra

In previous Sections we have encountered and discussed situations in which the energy spectrum is continuous, as for particles free to move far to infinity with or without potential barriers, and other cases presenting a discrete spectrum, like that of bound particles. We will now show that other different interesting situations exist, in particular those characterized by a band spectrum. That is the case for a particle in a periodic potential, like an electron in the atomic lattice of a solid.

An example which can be treated in a relatively simple way is that in which the potential energy can be written as the sum of an infinite number of thin barriers (Kronig-Penney model), each proportional to the Dirac delta function, placed at a constant distance a from each other:

$$V(x) = \sum_{n=-\infty}^{\infty} \mathcal{V}\delta(x - na). \tag{2.126}$$

It is clear that:

$$V(x + a) = V(x), \tag{2.127}$$

so that we are dealing with a periodic potential. Our analysis will be limited to the case of barriers, i.e. $\mathcal{V} > 0$.

Equation (2.127) expresses a symmetry property of the Schrödinger equation, which is completely analogous to the symmetry under axis reflection discussed for the square well and valid also in the case of the harmonic oscillator. With an argument similar to that used in the square well case, it can be shown that for periodic potentials, i.e. invariant under translations by a, if $\psi_E(x)$ is a solution of the stationary Schrödinger equation then $\psi_E(x + a)$ is a solution too, corresponding to the same energy, so that, by suitable linear combinations, the analysis can be limited to a particular class of functions which are not changed by the symmetry transformation but for an overall multiplicative constant. In the case of reflections that constant must be ± 1 since a double reflection must bring back to the original configuration. Instead, in the case of translations $x \to x + a$, solutions can be chosen so as to satisfy the following relation:

$$\psi_E(x + a) = \alpha\,\psi_E(x),$$

where α is in general a complex number. Clearly such functions, like plane waves, are not normalizable, so that we have to make reference to the collective physical interpretation, as in the case of the potential barrier. In this case probability densities which do not vanish in the limit $|x| \to \infty$ are acceptable, but those diverging in the same limit must be discarded anyway. That constrains α to be a pure phase factor, $\alpha = e^{i\phi}$, so that

$$\psi_E(x + a) = e^{i\phi}\,\psi_E(x). \tag{2.128}$$

This is therefore another application of the symmetry principle enunciated in Section 2.6.

The wave function $\psi_E(x)$ must satisfy both (2.128) and the free Schrödinger equation in each interval $(n - 1)a < x < na$:

$$-\frac{\hbar^2}{2m}\partial_x^2\psi_E(x) = E\,\psi_E(x),$$

which has the general solution

$$\psi_E(x) = a_n e^{i\sqrt{2mE}\,x/\hbar} + b_n e^{-i\sqrt{2mE}\,x/\hbar} \,.$$

Finally, at the position of each delta function, the wave function must be continuous while its first derivative must be discontinuous with a gap equal to $(2mV/\hbar^2)\psi_E(na)$. Since, according to (2.128), the wave function is pseudo-periodic, these conditions will be satisfied in every point $x = na$ if they are satisfied in the origin.

The continuity (discontinuity) conditions in the origin can be written as

$$a_0 + b_0 = a_1 + b_1 \,,$$

$$i\frac{\sqrt{2mE}}{\hbar}(a_1 - b_1 - a_0 + b_0) = \frac{2mV}{\hbar^2}(a_0 + b_0) \,, \qquad (2.129)$$

while (2.128), in the interval $-a < x < 0$, is equivalent to:

$$a_1 e^{i\sqrt{2mE}(x+a)/\hbar} + b_1 e^{-i\sqrt{2mE}(x+a)/\hbar} = e^{i\phi}\left(a_0 e^{i\sqrt{2mE}x/\hbar} + b_0 e^{-i\sqrt{2mE}x/\hbar}\right). \qquad (2.130)$$

Last equation implies:

$$a_1 = e^{i\left(\phi - \sqrt{2mE}a/\hbar\right)}a_0 \,, \qquad b_1 = e^{i\left(\phi + \sqrt{2mE}a/\hbar\right)}b_0 \,,$$

which replaced in (2.129) lead to a system of two homogeneous linear equation in two unknown quantities:

$$\left(1 - e^{i\left(\phi - \frac{\sqrt{2mE}}{\hbar}a\right)}\right) - i\sqrt{\frac{2m}{E}}\frac{V}{\hbar}\right)a_0 - \left(1 - e^{i\left(\phi + \frac{\sqrt{2mE}}{\hbar}a\right)}\right) + i\sqrt{\frac{2m}{E}}\frac{V}{\hbar}\right)b_0 = 0$$

$$\left(1 - e^{i\left(\phi - \frac{\sqrt{2mE}}{\hbar}a\right)}\right)a_0 + \left(1 - e^{i\left(\phi + \frac{\sqrt{2mE}}{\hbar}a\right)}\right)b_0 = 0 \,. \qquad (2.131)$$

The system admits non-trivial solutions $(a_0, b_0 \neq 0)$ if and only if the determinant of the coefficient matrix does vanish; that is equivalent to a second order equation for $e^{i\phi}$:

$$\left(1 - e^{i\left(\phi - \frac{\sqrt{2mE}}{\hbar}a\right)}\right) - i\sqrt{\frac{2m}{E}}\frac{V}{\hbar}\right)\left(1 - e^{i\left(\phi + \frac{\sqrt{2mE}}{\hbar}a\right)}\right)$$

$$+ \left(1 - e^{i\left(\phi + \frac{\sqrt{2mE}}{\hbar}a\right)}\right) + i\sqrt{\frac{2m}{E}}\frac{V}{\hbar}\right)\left(1 - e^{i\left(\phi - \frac{\sqrt{2mE}}{\hbar}a\right)}\right)$$

$$= 2e^{2i\phi} - \left(\left(2 - i\sqrt{\frac{2m}{E}}\frac{V}{\hbar}\right)e^{i\frac{\sqrt{2mE}}{\hbar}a} + \left(2 + i\sqrt{\frac{2m}{E}}\frac{V}{\hbar}\right)e^{-i\frac{\sqrt{2mE}}{\hbar}a}\right)e^{i\phi}$$

$$+ 2 = 0 \,, \qquad (2.132)$$

which can be rewritten in the form:

$$\mathrm{e}^{2\mathrm{i}\phi} - \left(2\cos\left(\frac{\sqrt{2mE}}{\hbar}a\right) + \sqrt{\frac{2m}{E}\frac{\mathcal{V}}{\hbar}}\sin\left(\frac{\sqrt{2mE}}{\hbar}a\right)\right)\mathrm{e}^{\mathrm{i}\phi} + 1$$

$$\equiv \mathrm{e}^{2\mathrm{i}\phi} - 2A\mathrm{e}^{\mathrm{i}\phi} + 1 = 0. \tag{2.133}$$

Equation (2.133) can be solved by a real ϕ if and only if $A^2 < 1$, as can be immediately verified by using the resolutive formula for second degree equations.

We have therefore an inequality, involving the energy E together with the amplitude $\mathcal{V}$ and the period a of the potential, which is a necessary and sufficient condition for the existence of physically acceptable solutions of the Schrödinger equation:

$$\left(\cos\left(\frac{\sqrt{2mE}}{\hbar}a\right) + \sqrt{\frac{m}{2E}\frac{\mathcal{V}}{\hbar}}\sin\left(\frac{\sqrt{2mE}}{\hbar}a\right)\right)^2 < 1, \tag{2.134}$$

hence

$$\cos^2\left(\frac{\sqrt{2mE}}{\hbar}a\right) + \frac{m}{2E}\frac{\mathcal{V}^2}{\hbar^2}\sin^2\left(\frac{\sqrt{2mE}}{\hbar}a\right)$$

$$+2\sqrt{\frac{m}{2E}}\frac{\mathcal{V}}{\hbar}\sin\left(\frac{\sqrt{2mE}}{\hbar}a\right)\cos\left(\frac{\sqrt{2mE}}{\hbar}a\right) < 1 \tag{2.135}$$

and therefore

$$1 - \cos^2\left(\frac{\sqrt{2mE}}{\hbar}a\right) - \frac{m}{2E}\frac{\mathcal{V}^2}{\hbar^2}\sin^2\left(\frac{\sqrt{2mE}}{\hbar}a\right)$$

$$-2\sqrt{\frac{m}{2E}}\frac{\mathcal{V}}{\hbar}\sin\left(\frac{\sqrt{2mE}}{\hbar}a\right)\cos\left(\frac{\sqrt{2mE}}{\hbar}a\right)$$

$$= \left(1 - \frac{m}{2E}\frac{\mathcal{V}^2}{\hbar^2}\right)\sin^2\left(\frac{\sqrt{2mE}}{\hbar}a\right)$$

$$-2\sqrt{\frac{m}{2E}}\frac{\mathcal{V}}{\hbar}\sin\left(\frac{\sqrt{2mE}}{\hbar}a\right)\cos\left(\frac{\sqrt{2mE}}{\hbar}a\right) < 0, \tag{2.136}$$

leading finally to:

$$\cot\left(\frac{\sqrt{2mE}}{\hbar}a\right) < \frac{1}{2}\left(\sqrt{\frac{2E}{m}\frac{\hbar}{\mathcal{V}}} - \sqrt{\frac{m}{2E}\frac{\mathcal{V}}{\hbar}}\right). \tag{2.137}$$

Both sides of last inequality are plotted in Fig. 2.5 for a particular choice of the parameter $\gamma = \hbar^2/(ma\mathcal{V}) = 1/4$. The variable used in the figure is $x = \sqrt{2mE}a/\hbar$, so that the two plotted functions are $f_1 = \cot x$ and

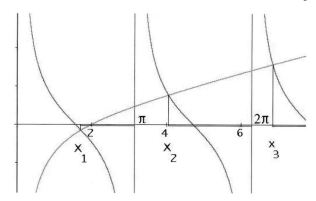

Fig. 2.5. The plot of inequality (2.137) identifying the first three bands of al-
lowed values of $\sqrt{2mE}\, a/\hbar$ for the choice of the Kronig–Penney parameter $\gamma = \hbar^2/(ma\mathcal{V}) = 1/4$

$f_2 = (\gamma x - 1/(\gamma x))/2$. The intervals where the inequality (2.137) is satisfied
are those enclosed between x_1 and π, x_2 and 2π, x_3 and 3π and so on. Indeed
in these regions the uniformly increasing function f_2 is greater than the oscillating function f_1. The result shows therefore that the permitted energies
correspond to a series of intervals $(x_n, n\pi)$, which are called *bands*, separated
by a series of forbidden *gaps*.

As we will discuss in the next Chapter, electrons in a solid, which are
compelled by the *Pauli exclusion principle* to occupy each a different energy
level, may fill completely a certain number of bands, so that they can only
absorb energies greater than a given minimum quantity, corresponding to the
gap with the next free band: in such situation electrons behave as bound
particles. Alternatively, if the electrons fill partially a given band, they can
absorb arbitrarily small energies, thus behaving as free particles. In the first
case the solid is an *insulator*, in the second it is a *conductor*.

Having determined the phase $\phi(E)$ from (2.133) and taking into account
(2.128), it can be seen that, by a simple transformation of the wave function:

$$\psi_E(x) \equiv e^{i\,\phi(E)\,x/a}\hat{\psi}_E(x) \equiv e^{\pm i\,p(E)\,x/\hbar}\hat{\psi}_E(x)\,, \qquad (2.138)$$

equation (2.128) can be translated into a periodicity constraint:

$$\hat{\psi}_E(x + a) = \hat{\psi}_E(x)\,.$$

Therefore wave functions in a periodic potential can be written as in (2.138),
i.e. like plane waves, which are called *Bloch waves*, modulated by periodic
functions $\hat{\psi}_E(x)$.

It must be noticed that the momentum associated with Bloch waves,
$p(E) = (\hbar/a)\phi(E)$, cannot take all possible real values, as in the case of
free particles, but is limited to the interval $(-\hbar\pi/a,\ \hbar\pi/a)$, which is known

as the *first Brillouin zone*. This limitation can be seen as the mathematical reason underlying the presence of bands.

On the other hand the relation which in a given band gives the electron energy as a function of the Bloch momentum (dispersion relation) is very intricate from the analytical point of view. It is indeed the inverse function of $p(E) = (\hbar/a)\arccos(A(E))$ with $A(E)$ defined by (2.133). For that reason we limit ourselves to some qualitative remarks.

By noticing that in the lower ends of the bands, x_n, $n = 1, 2, ..$, the parameter A in (2.133) is equal to

$$\cos x_n + \frac{\sin x_n}{\gamma \, x_n} = \frac{\sin x_n}{2}\left(\gamma \, x_n + \frac{1}{\gamma \, x_n}\right) = (-1)^{n+1} , \qquad (2.139)$$

we have: $e^{i\phi}|_{x_n} = (-1)^{n+1}$. Hence $\phi(x_n) = 0$ for odd n and $\phi(x_n) = \pm\pi$ for even n. Instead in the upper ends, $x = n\pi$, we have $A = \cos n\pi = (-1)^n$ hence $\phi(n\pi) = 0$ for even n and $\phi(n\pi) = \pm\pi$ for odd n. Moreover, for a generic A between -1 and 1, there are two solutions: $A \pm i\sqrt{1 - A^2}$ corresponding to opposite phases ($\phi(E) = \pm\arctan(\sqrt{1 - A^2}/A)$) interpolating between 0 and $\pm\pi$.

Therefore, based on Fig. 2.5, we come to the conclusion that in odd bands the minimum energy corresponds to states with $p = 0$, while states at the border of the Brillouin zone have the maximum possible energy. The opposite happens instead for even bands. Finally we observe that the derivative dE/dp vanishes at the border of the Brillouin zone, where $A^2 = 1$ and A has a non-vanishing derivative, indeed we have

$$\frac{dE}{dp} = \frac{a}{\hbar}\left(\frac{d(\arccos A(E))}{dE}\right)^{-1} = \pm\frac{a}{\hbar}\frac{\sqrt{1 - A^2}}{dA(E)/dE} . \qquad (2.140)$$

Suggestions for Supplementary Readings

- E. H. Wichman: *Quantum Physics - Berkeley Physics Course*, volume 4 (Mcgraw-Hill Book Company, New York 1971)
- L. D. Landau, E. M. Lifchitz: *Quantum Mechanics - Non-relativistic Theory*, Course of Theoretical Physics, volume 3 (Pergamon Press, London 1958)
- L. I. Schiff: *Quantum Mechanics* 3d edn (Mcgraw-Hill Book Company, Singapore 1968)
- J. J. Sakurai: *Modern Quantum Mechanics* (The Benjamin-Cummings Publishing Company Inc., Menlo Park 1985)
- E. Persico: *Fundamentals of Quantum Mechanics* (Prentice - Hall Inc., Englewood Cliffs 1950)

Problems

2.1. A biatomic molecule can be simply described as two point-like objects of mass $m = 10^{-26}$ Kg placed at a fixed distance $d = 10^{-9}$ m. Describe what are the possible values of the molecule energy according to Bohr's quantization rule. Compute the energy of the photons which are emitted when the system decays from the $(n + 1)$-th to the n-th energy level.

Answer: $E_{n+1} - E_n = (\hbar^2/2I)(n + 1)^2 - (\hbar^2/2I)n^2 = (2n + 1)\hbar^2/(md^2) \simeq 1.1 \ 10^{-24}(2n + 1)$ J. Notice that in Sommerfeld's perfected theory, mentioned in Section 2.2, the energy of a rotator is given by $E_n = \hbar^2 n(n + 1)/2I$, so that the factor $2n + 1$ in the solution must be replaced by $2n + 2$.

2.2. An artificial satellite of mass $m = 1$ Kg rotates around the earth along a circular orbit of radius practically equal to that of the earth itself, i.e. roughly 6370 Km. If the satellite orbits are quantized according to Bohr's rule, how much does the radius change when going from one quantized level to the next (i.e. from n to $n + 1$)?

Answer: If g indicates the gravitational acceleration at the earth surface, the radius of the n-th orbit is given by $r_n = (n^2\hbar^2/m^2R^2g)$. Therefore, if $r_n = R$, $\delta r_n \equiv r_{n+1} - r_n \simeq 2\hbar/m\sqrt{Rg} \simeq 2.6 \ 10^{-38}$ m .

2.3. An electron is accelerated through a potential difference $\Delta V = 10^8$ V, what is its wavelength according to de Broglie?

Answer: The energy gained by the electron is much greater than mc^2, therefore it is ultra-relativistic and its momentum is $p \simeq E/c$. Hence $\lambda \simeq hc/e\Delta V \simeq 2 \ 10^{-15}$ m. The exact formula is instead $\lambda = hc/\sqrt{(e\Delta V + mc^2)^2 - m^2c^4}$.

2.4. An electron is constrained to bounce between two reflecting walls placed at a distance $d = 10^{-9}$ m from each other. Assuming that, as in the case of a stationary electromagnetic wave confined between two parallel mirrors, the distance d be equal to n half wavelengths, determine the possible values of the electron energy as a function of n.

Answer: $E_n = \hbar^2\pi^2 n^2/(2md^2) \simeq n^2 \ 6.03 \ 10^{-20}$ J .

2.5. An electron of kinetic energy 1 eV is moving upwards under the action of its weight. Can it reach an altitude of 1 Km? If yes, what is the variation of its de Broglie wavelength?

Answer: The maximum altitude reachable by the electron in a constant gravitational field would be $h = T/mg \simeq 1.6 \ 10^{10}$ m. After one kilometer the kinetic energy changes by $\delta T/T \simeq 5.6 \ 10^{-8}$, hence $\delta\lambda/\lambda \simeq 2.8 \ 10^{-8}$. Since the starting wavelength is $\lambda = h/\sqrt{2mT} \simeq 1.2 \ 10^{-9}$ m, the variation is $\delta\lambda \simeq 3.4 \ 10^{-17}$ m.

2.6. Ozone (O_3) is a triatomic molecule made up of three atoms of mass $m \simeq 2.67\ 10^{-26}$ Kg placed at the vertices of an equilateral triangle with sides of length l. The molecule can rotate around an axis P going through its center of mass and orthogonal to the triangle plane, or around another axis L which passes through the center of mass as well, but is orthogonal to the first axis. Making use of Bohr's quantization rule and setting $l = 10^{-10}$ m, compare the possible rotational energies in the two different cases of rotations around P or L.

Answer: The moments of inertia are $I_P = ml^2 = 2.67\ 10^{-46}$ Kg m^2 and $I_L = ml^2/2 = 1.34\ 10^{-46}$ Kg m^2. The rotational energies are then $E_{L,n} = 2E_{P,n} = \hbar^2 n^2/2I_L \simeq n^2\ 4.2\ 10^{-23}$ J .

2.7. A table-salt crystal is irradiated with an X-ray beam of wavelength $\lambda = 2.5\ 10^{-10}$ m, the first diffraction peak $(d\sin\theta = \lambda)$ is observed at an angle equal to $26.3°$. What is the interatomic distance of salt?

Answer: $d = \lambda/\sin\theta \simeq 5.6\ 10^{-10}$ m .

2.8. In β decay a nucleus, with a radius of the order of $R = 10^{-14}$ m, emits an electron with a kinetic energy of the order of 1 MeV $= 10^6$ eV. Compare this value with that which according to the uncertainty principle is typical of an electron initially localized inside the nucleus (thus having a momentum $p \sim \hbar/R$).

Answer: The order of magnitude of the momentum of the particle is $p \sim \hbar/R \sim 10^{-20}$ N s, thus $pc \simeq 3\ 10^{-12}$ J, which is much larger that the electron rest energy $m_e c^2 \simeq 8\ 10^{-14}$ J. Therefore the kinetic energy of the electron in the nucleus is about $pc = 3.15\ 10^{-12}$ J $\simeq 20$ MeV.

2.9. An electron is placed in a constant electric field $\mathcal{E} = 1000$ V/m directed along the x axis and going out of a plane surface orthogonal to the same axis. The surface also acts on the electron as a reflecting plane where the electron potential energy $V(x)$ goes to infinity. The behaviour of $V(x)$ is therefore as illustrated in the figure:

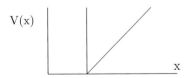

Evaluate the order of magnitude of the minimal electron energy according to Heisenberg's Uncertainty Principle.

Answer: The total energy is given by $\epsilon = p^2/2m + V(x) = p^2/2m + e\mathcal{E}x$, with the constraint $x > 0$. From a classical point of view, the minimal energy would be realized for a particle at rest $(p = 0)$ in the minimum of the potential. The uncertainty principle states instead that $\delta p\,\delta x \sim \hbar$, where δx is the size of a region around the potential minimum where the electron is localized. Therefore the minimal total energy compatible with the uncertainty principle can be written as a function of δx as $E(\delta x) \equiv \hbar^2/(2m\delta x^2) + e\mathcal{E}\delta x$ $(\delta x > 0)$ and has a minimum

$$\epsilon_{min} \sim \frac{3}{2}\left(\frac{\hbar^2 e^2 \mathcal{E}^2}{m}\right)^{1/3} \sim 0.610^{-4}\ \text{eV}.$$

2.10. A KBr molecule can be roughly described as a system of two particles of opposite charge and mass $M_K = 6.5\ 10^{-26}$ Kg and $M_{Br} = 13\ 10^{-26}$ Kg, which are kept at a fixed distance $d = 10^{-9}$ m by interatomic forces. By applying Bohr's quantization rule to the angular momentum of the system computed with respect to its center of mass, determine the frequency spectrum of the photons emitted by the molecule in transitions from level n to level $(n-1)$.

Answer: Bohr's rotational energy levels are given by $E_n = n^2\hbar^2/(2I)$, where $I = d^2 M_K M_{Br}/(M_K + M_{Br})$. The photons emitted in the indicated transitions have energy $h\nu_{n\to n-1} = E_n - E_{n-1} = (2n-1)\hbar^2/2I$, hence $\nu_{n\to n-1} \simeq (2n-1)1.96\ 10^8$ Hz.

2.11. An atom of mass $M = 10^{-26}$ Kg is attracted towards a fixed point by an elastic force of constant $k = 1$ N/m; the atom is moving along a circular orbit in a plane orthogonal to the x axis. Determine the energy levels of the system by making use of Bohr's quantization rule for the angular momentum computed with respect to the fixed point.

Answer: Let ω be the angular velocity and r the orbital radius. The centripetal force is equal to the elastic one, hence $\omega = \sqrt{k/M}$. The total energy is given by $E = (1/2)M\omega^2 r^2 + (1/2)kr^2 = M\omega^2 r^2 = L\omega$, where $L = M\omega r^2$ is the angular momentum. Since $L = n\hbar$, we finally infer $E_n = n\hbar\omega = n\ 10^{-21}$ J $\simeq n\ 0.66\ 10^{-2}$ eV.

2.12. Compute the number of photons emitted in one second by a lamp of power 10 W, if the photon wavelength is $0.5\ 10^{-6}$ m.

Answer: The energy of a single photon is $E = h\nu$ and $\nu = c/\lambda = 6\ 10^{14}$ Hz, hence $E = 4\ 10^{-19}$ J. Therefore the number photons emitted in one second is $2.5\ 10^{19}$.

2.13. A particle of mass $m = 10^{-28}$ Kg is moving along the x axis under the influence of a potential energy given by $V(x) = v\sqrt{|x|}$, where $v = 10^{-15}$ J m$^{-1/2}$. Determine what is the order of magnitude of the minimal electron energy according to the uncertainty principle.

Answer: The total energy of the particle is given by

$$E = \frac{p^2}{2m} + v\sqrt{|x|}.$$

If the particle is localized in a region of size δx around the minimum of the potential ($x = 0$), according to the uncertainty principle it has a momentum at least of the order of $\delta p = \hbar/\delta x$. It is therefore necessary to minimize the quantity

$$E = \frac{\hbar^2}{2m\delta x^2} + v\sqrt{\delta x}$$

with respect to δx, finally finding that

$$E_{min} \simeq \left(\frac{\hbar^2 v^4}{m}\right)^{1/5} \left(2^{1/5} + 2^{-9/5}\right) \simeq 0.092 \text{ eV}.$$

It is important to notice that our result, apart from a numerical factor, could be also predicted on the basis of simple dimensional remarks. Indeed, it can be easily checked that $(\hbar^2 v^4/m)^{1/5}$ is the only possible quantity having the dimensions of an energy and constructed in terms of m, v and $\hbar$, which are the only physical constants involved in the problem. In the analogous classical problem $\hbar$ is missing, and v and m are not enough to build a quantity with the dimensions of energy, hence the classical problem lacks the typical energy scale appearing at the quantum level.

2.14. An electron beam with kinetic energy equal to 10 eV is split into two parallel beams placed at different altitudes in the terrestrial gravitational field. If the altitude gap is $d = 10$ cm and if the beams recombine after a path of length L, say for which values of L the phases of the recombining beams are opposite (destructive interference). Assume that the upper beam maintains its kinetic energy , that the total energy is conserved for both beams and that the total phase difference accumulated during the splitting and the recombination of the beams is negligible.

Answer: De Broglie's wave describing the initial electron beam is proportional to $\exp(\mathrm{i}px/\hbar - \mathrm{i}Et/\hbar)$, where $p = \sqrt{2mE_k}$ is the momentum corresponding to a kinetic energy E_k and $E = E_k + mgh$ is the total energy. The beam is split into two beams, the first travels at the same altitude and is described by the same wave function, the second travels 10 cm lower and is described by a wave function $\propto \exp(\mathrm{i}p'x/\hbar - \mathrm{i}Et/\hbar)$ where $p' = \sqrt{2mE_k'} = \sqrt{2m(E_k + mgd)}$ (obviously the total energy E stays unchanged). The values of L for which the two beams recombine with opposite phases are solutions of $(p' - p)L/\hbar = (2n + 1)\pi$ where n is an integer. The smallest value of L is $L = \pi\hbar/(p' - p)$. Notice that $mgd \simeq 10^{-30}$ $J \ll 10$ eV $\simeq 1.6 \ 10^{-18}$ J hence $p' - p \simeq \sqrt{2mE_k}(mgd/2E_k)$ and $L \simeq 2\pi\hbar E_k/(mgd\sqrt{2mE_k}) \simeq 700$ m. This experiment, which clearly demonstrates the wavelike behaviour of material particles, has been really performed but using neutrons in place of electrons: the use of neutrons has various advantages, among which that of leading to smaller values of L because of the much heavier mass, as it is clear from the solution. That makes the setting of the experimental apparatus simpler.

2.15. An electron is moving in the $x - y$ plane under the influence of a magnetic induction field parallel to the z-axis. What are the possible energy levels according to Bohr's quantization rule?

Answer: The electron is subject to the force $e\boldsymbol{v} \wedge \boldsymbol{B}$ where $\boldsymbol{v}$ is its velocity. Classically the particle, being constrained in the $x - y$ plane, would move on circular orbits with constant angular velocity $\omega = eB/m$, energy $E = 1/2\ m\omega^2 r^2$ and any radius r. Bohr's quantization instead limits the possible values of r by $m\omega r^2 = n\hbar$. Finally one finds $E = 1/2\ n\hbar\omega = n\hbar eB/(2m)$. This is an approximation of the exact solution for the quantum problem of an electron in a magnetic field (Landau's levels).

2.16. The positron is a particle identical to the electron but carrying an opposite electric charge. It forms a bound state with the electron, which is similar to the hydrogen atom but with the positron in place of the proton: that is called *positronium*. What are its energy levels according to Bohr's rule?

Answer: The computation goes exactly along the same lines as for the proton–electron system, but the reduced mass $\mu = m^2/(m + m) = m/2$ has to be used in place of the electron mass. energy levels are thus

$$E_n = -\frac{me^4}{16\epsilon_0\hbar^2 n^2} \ .$$

2.17. A particle of mass $M = 10^{-29}$ Kg is moving in two dimensions under the influence of a central potential

$$V = \sigma r \ ,$$

where $\sigma = 10^5$ N. Considering only circular orbits, what are the possible values of the energy according to Bohr's quantization rule?

Answer: Combining the equation for the centripetal force necessary to sustain the circular motion, $m\omega^2 r = \sigma$, with the quantization of angular momentum, $m\omega r^2 = n\hbar$, we obtain for the total energy, $E = 1/2\ m\omega^2 r^2 + \sigma r$, the following quantized values

$$E_n = \frac{3}{2}\left(\frac{\hbar^2\sigma^2}{m}\right)^{1/3} n^{2/3} \simeq 2\ n^{2/3} \text{ GeV} \ .$$

Notice that the only possible combination of the physical parameters available in the problem with energy dimensions is $(\hbar^2\sigma^2/m)^{1/3}$. The potential proposed in this problem is similar to that believed to act among quarks, which are the elementary constituent of hadrons (a wide family of particles including protons, neutrons, mesons ...); σ is usually known as the *string tension*. Notice that the lowest energy coincides, identifying $\sigma = e\mathcal{E}$, with that obtained in Problem 2.9 using the uncertainty principle for the one-dimensional problem.

2.18. The momentum probability distribution for a particle with wave function $\psi(x)$ is given by

$$\left|\int_{-\infty}^{\infty} dx \frac{1}{\sqrt{h}} e^{ipx/\hbar}\psi(x)\right|^2 \equiv |\tilde{\psi}(p)|^2 \ .$$

Compute that distribution for the following wave function $\psi(x) = e^{-a|x|/2}\sqrt{a/2}$ (a is real and positive) and verify the validity of the uncertainty principle in this case.

Answer: $\tilde{\psi}(p) = (\hbar a)^{3/2}/(\sqrt{4\pi}(p^2 + \hbar^2 a^2/4))$ hence

$$\Delta_x^2 = \frac{a}{2}\int_{-\infty}^{\infty} dx\, x^2 e^{-a|x|} = \frac{2}{a^2}\,,$$

$$\Delta_p^2 = \frac{(\hbar a)^3}{4\pi}\int_{-\infty}^{\infty} dp\, \frac{p^2}{(p^2 + \frac{a^2\hbar^2}{4})^2} = \frac{a^2\hbar^2}{4}\,,$$

so that $\Delta_x^2\Delta_p^2 = \hbar^2/2 > \hbar^2/4$.

2.19. The wave function of a free particle is

$$\psi(x) = \frac{1}{\sqrt{2P\hbar}}\int_{-P}^{P} dq\, e^{\frac{iqx}{\hbar}}$$

at time $t = 0$. What is the corresponding probability density $\rho(x)$ of locating the particle in a given point x? Give an integral representation of the wave function at a generic time t, assuming that the particle mass is m.

Answer: The probability density is $\rho(x) = |\psi(x)|^2 = \hbar/(\pi Px^2)\sin^2(Px/\hbar)$ while $\psi(x,t) = (1/\sqrt{2P\hbar})\int_{-P}^{P} dq\, e^{i(qx - q^2t/2m)/\hbar}$.

2.20. An electron beam hits the potential step sketched in the figure, coming from the right. In particular, the potential energy of the electrons is 0 for $x < 0$ and $-\mathcal{V} = -300$ eV for $x > 0$, while their kinetic energy in the original beam (thus for $x > 0$) is $E_k = 400$ eV. What is the reflection coefficient?

Answer: The wave function can be written, leaving aside an overall normalization coefficient which is not relevant for computing the reflection coefficient, as

$$\psi(x) = be^{-ik'x} \text{ for } x < 0\,, \qquad\qquad \psi(x) = e^{-ikx} + ae^{ikx} \text{ for } x > 0$$

where $k = \sqrt{2mE_k}/\hbar = \sqrt{2m(E+\mathcal{V})}/\hbar$ and $k' = \sqrt{2m(E_k - \mathcal{V})}/\hbar = \sqrt{2mE}/\hbar$; m is the electron mass and $E = E_k - \mathcal{V}$ is the total energy of the electrons. The continuity conditions at the position of the step read

$$b = 1 + a\,, \qquad\qquad bk' = (1 - a)k\,,$$

hence

$$b = \frac{2}{1 + k'/k} \, , \qquad\qquad a = \frac{1 - k'/k}{1 + k'/k} \, ,$$

and

$$R = |a|^2 = \left(\frac{k - k'}{k + k'}\right)^2 = \frac{2E + \mathcal{V} - 2\sqrt{E(E + \mathcal{V})}}{2E + \mathcal{V} + 2\sqrt{E(E + \mathcal{V})}} = \frac{1}{9} \, .$$

2.21. An electron beam hits the same potential step considered in Problem 2.20, this time coming from the left with a kinetic energy $E = 100$ eV. What is the reflection coefficient in this case?

Answer: In this case we write:

$$\psi(x) = e^{ik'x} + be^{-ik'x} \quad \text{for} \ \ x < 0 \, , \qquad\qquad \psi(x) = ae^{ikx} \quad \text{for} \ \ x > 0 \, ,$$

where again $k = \sqrt{2m(E + \mathcal{V})}/\hbar$ and $k' = \sqrt{2mE}/\hbar$ with $E = E_k$ being the total energy. By solving the continuity conditions we find:

$$b = \frac{k'/k - 1}{1 + k'/k} \ ; \quad R = |b|^2 = \left(\frac{k' - k}{k + k'}\right)^2 = \frac{2E + \mathcal{V} - 2\sqrt{E(E + \mathcal{V})}}{2E + \mathcal{V} + 2\sqrt{E(E + \mathcal{V})}} = \frac{1}{9} \, .$$

We would like to stress that the reflection coefficient coincides with that obtained in Problem 2.20: electron beams hitting the potential step from the right or from the left are reflected in exactly the same way, if their total energy E is the same, as it is in the present case. In fact this is a general result which is valid for every kind of potential barrier and derives directly from the invariance of the Schrödinger equation under time reversal: the complex conjugate of a solution is also a solution. It may seem a striking result, but it should not be so striking for those familiar with reflection of electromagnetic signals in presence of unmatching impedances.

Notice also that there is actually a difference between the two cases, consisting in a different sign for the reflected wave. That is irrelevant for computing R but significant for considering interference effects involving the incident and the reflected waves. In the present case interference is destructive, hence the probability density is suppressed close to the step, while in Problem 2.20 the opposite happens. To better appreciate this fact consider the analogy with an oscillating rope made up of two different ropes having different densities (which is a system in some sense similar to ours), and try to imagine the different behaviors observed if you enforce oscillations shaking the rope from the heavier (right-hand in our case) or from the lighter (left-hand in our case) side. As extreme and easier cases you could think of a single rope with a free end (one of the densities goes to zero) or with a fixed end (one of the densities goes to infinity): the shape of the rope at the considered endpoint is cosine-like in the first case and sine-like in the second case, exactly as for the cases of respectively the previous and the present problem in the limit $\mathcal{V} \to \infty$.

2.22. An electron beam hits, coming from the right, a potential step similar to that considered in Problem 2.20. However this time $-\mathcal{V} = -10$ eV and the kinetic energy of the incoming electrons is $E_k = 9$ eV. If the incident current is equal to $J = 10^{-3}$ A, compute how many electrons can be found, at a given time, along the negative x axis, i.e. how many electrons penetrate the step barrier reaching positions which would be classically forbidden.

Answer: The solution of the Schrödinger equation can be written as

$$\psi(x) = c\, e^{p'x/\hbar} \quad \text{for} \quad x < 0 , \qquad \psi(x) = a\, e^{ipx/\hbar} + b\, e^{-ipx/\hbar} \quad \text{for} \quad x > 0$$

where $p = \sqrt{2mE}$ and $p' = \sqrt{2m(V-E)}$. Imposing continuity in $x = 0$ for both $\psi(x)$ and its derivative, we obtain $c = 2a/(1 + ip'/p)$ and $b = a(1 - ip'/p)/(1 + ip'/p)$. It is evident that $|b|^2 = |a|^2$, hence the reflection coefficient is one. Indeed the probability current $J(x) = -i\hbar/(2m)(\psi^*\partial_x\psi - \psi\partial_x\psi^*)$ vanishes on the left, where we have an evanescent wave function, hence no transmission. Nevertheless the probability distribution is non-vanishing for $x < 0$ and, on the basis of the collective interpretation, the total number of electrons on the left is given by

$$N = \int_{-\infty}^{0} |\psi(x)|^2 dx = |c|^2 \hbar/(2p') = \frac{2|a|^2 \hbar p^2}{p'(p^2 + p'^2)}.$$

The coefficient a can be computed by asking that the incident current $J_{el} = eJ = e|a|^2 p/m \equiv 10^{-3}$ A. The final result is $N \simeq 1.2$.

2.23. An electron is confined inside a cubic box with reflecting walls and an edge of length $L = 2\ 10^{-9}$ m. How many levels have energy less than 1 eV? Take into account the spin degree of freedom, which in practice doubles the number of available levels.

Answer: Energy levels in a cubic box are $E_{n_x,n_y,n_z} = \pi^2\hbar^2(n_x^2 + n_y^2 + n_z^2)/(2mL^2)$, where $m = 0.911\ 10^{-30}$ Kg and n_x, n_y, n_z are positive integers. The constraint $E < 1$ eV implies $n_x^2 + n_y^2 + n_z^2 < 10.2$, which is satisfied by 7 different combinations $((1,1,1),\ (2,1,1),\ (2,2,1)$ plus all possible different permutations). Taking spin into account, the number of available levels is 14.

2.24. When a particle beam hits a potential barrier and is partially transmitted, a forward going wave is present on the other side of the barrier which, besides having a reduced amplitude with respect to the incident wave, has also acquired a phase factor which can be inferred by the ratio of the transmitted wave coefficient to that of the incident one. Assuming a thin barrier describable as

$$V(x) = v\,\delta(x) ,$$

and that the particles be electrons of energy $E = 10$ eV, compute the value of v for which the phase difference is $-\pi/4$.

Suppose now that we have two beams of equal amplitude and phase and that one beam goes through the barrier while the other goes free. The two beams recombine after paths of equal length. What is the ratio of the recombined beam intensity to that one would have without the barrier?

Answer: On one side of the barrier the wave function is $e^{i\sqrt{2mE}x/\hbar} + a\, e^{-i\sqrt{2mE}x/\hbar}$, while it is $b\, e^{i\sqrt{2mE}x/\hbar}$ on the other side. Continuity and discontinuity constraints read $1 + a = b$ and $b - 1 + a = \sqrt{2m/E}vb/\hbar$, from which $b = (1 + i\sqrt{m/2E}v/\hbar)^{-1}$

can be easily derived. Requiring that the phase of b be $-\pi/4$ is equivalent to $\sqrt{m/2Ev}/\hbar = 1$, hence $v \simeq 2.0\ 10^{-28}$ J m.

With this choice of v the recombined beam is $[1 + 1/(1+i)]e^{i\sqrt{2mEx}/\hbar}$. The ratio of the intensity of the recombined beam to that one would have without the barrier is $\|[1 + 1/(1+i)]/2\|^2 = 5/8$.

2.25. If a potential well in one dimension is so thin as to be describable by a Dirac delta function:
$$V(x) = -VL\delta(x)$$
where V is the depth and L the width of the well, then it is possible to compute the bound state energies by recalling that for a potential energy of that kind the wave function is continuous while its first derivative has the following discontinuity:
$$\lim_{\epsilon \to 0} (\partial_x \psi(x + \epsilon) - \partial_x \psi(x - \epsilon)) = -\frac{2m}{\hbar^2} VL\psi(0) .$$

What are the possible energy levels?

Answer: The bound state wave function is
$$a\, e^{-\sqrt{2mBx}/\hbar} \text{ for } x > 0 \qquad \text{and} \qquad a\, e^{\sqrt{2mBx}/\hbar} \text{ for } x < 0$$
where the continuity condition for the wave function has been already imposed. $B = |E|$ is the absolute value of the energy (which is negative for a bound state). The discontinuity condition on the first derivative leads to an equation for B which has only one solution, $B = mV^2L^2/(2\hbar^2)$, thus indicating the existence of a single bound state.

2.26. A particle of mass m moves in the following one dimensional potential:
$$V(x) = v(\alpha\delta(x - L) + \alpha\delta(x + L) - \frac{1}{L}\theta(L^2 - x^2)) ,$$
where $\theta(y) = 0$ for $y < 0$ and $\theta(y) = 1$ for $y > 0$. Constants are such that
$$\frac{2mvL}{\hbar^2} = \left(\frac{\pi}{4}\right)^2 .$$

For what value of $\alpha > 0$ are there any bound states?

Answer: The potential is such that $V(-x) = V(x)$: in this case the lowest energy level, if any, corresponds to an even wave function. We can thus limit the discussion to the region $x > 0$, where we have $\psi(x) = \cos kx$ for $x < L$ and $\psi(x) = ae^{-\beta x}$ for $x > L$, with the constraint $0 < k < \sqrt{2mv/(\hbar^2 L)} = \pi/(4L)$, since $\beta = \sqrt{2mv/(\hbar^2 L) - k^2}$ must be real in order to have a bound state, hence $kL < \pi/4$. Continuity and discontinuity constraints, respectively on $\psi(x)$ and $\psi'(x)$ in $x = L$, give: $\tan kL = (\beta + 2mv\alpha/\hbar^2)/k$. Setting $y \equiv kL$, we have

$$\tan y = \frac{\sqrt{\frac{\pi^2}{16} - y^2 + \frac{\pi^2}{16}\alpha}}{y}.$$

The function on the left hand side grows from 0 to 1 in the interval $0 < y < \pi/4$, while the function on the right decreases from ∞ to $\pi/4\alpha$ in the same interval. Therefore an intersection (hence a bound state) exists only if $\alpha < 4/\pi$.

2.27. An electron moves in a one dimensional potential corresponding to a square well of depth $V = 0.1$ eV and width $L = 3 \ 10^{-10}$ m. Show that in these conditions there is only one bound state and compute its energy in the thin well approximation, discussing also the validity of that limit.

Answer: There is one only bound state if the first odd state is absent. That is true if $y = \sqrt{2mVL}/(2\hbar) < \pi/2$, which is verified in our case since, using $m = 0.911 \ 10^{-30}$ Kg, one obtains $y \simeq 0.243 < \pi/2$.

Setting $B \equiv -E$, where E is the negative energy of the bound state, B is obtained as a solution of

$$\tan\left(\frac{L\sqrt{2m(V - B)}}{2\hbar}\right) = \sqrt{\frac{B}{V - B}}.$$

The thin well limit corresponds to $V \to \infty$ and $L \to 0$ as the product VL is kept constant. Neglecting B with respect to V we can write

$$\tan\left(\frac{\sqrt{2mVL^2}}{2\hbar}\right) = \sqrt{\frac{B}{V}}.$$

In the thin well limit $VL^2 \to \text{constant} \cdot L \to 0$, hence we can replace the tangent by its argument, obtaining finally $B = mV^2L^2/(2\hbar^2) \simeq 0.59 \ 10^{-2}$ eV, which coincides with the result obtained in Problem 2.25. In this case the argument of the tangent is $y \sim 0.24$ and we have $\tan 0.24 \simeq 0.245$; therefore the exact result differs from that obtained in the thin well approximation by roughly 4 %.

2.28. An electron moves in one dimension and is subject to forces corresponding to a potential energy:

$$V(x) = \mathcal{V}[-\delta(x) + \delta(x - L)].$$

What are the conditions for the existence of a bound state and what is its energy if $L = 10^{-9}$ m and $\mathcal{V} = 2 \ 10^{-29}$ J m ?

Answer: A solution of the Schrödinger equation corresponding to a binding energy $B \equiv -E$ can be written as

$$\psi(x) = e^{\sqrt{2mBx}/\hbar} \quad \text{for} \quad x < 0,$$

$$\psi(x) = ae^{\sqrt{2mBx}/\hbar} + be^{-\sqrt{2mBx}/\hbar} \quad \text{for} \quad 0 < x < L,$$

$$\psi(x) = ce^{-\sqrt{2mBx}/\hbar} \quad \text{for} \quad L < x.$$

Continuity and discontinuity constraints, respectively for the wave function and for its derivative in $x = 0$ and $x = L$, give: $a + b = 1$, $a - b - 1 = -\sqrt{2m/B\mathcal{V}}/\hbar$, $ae^{\sqrt{8mBL}/\hbar} + b = c$, $ae^{\sqrt{8mBL}/\hbar} - b + c = -c\sqrt{2m/B\mathcal{V}}/\hbar$.

The four equations are compatible if $e^{\sqrt{8mBL}/\hbar} = (1 - \frac{2B\hbar^2}{m\mathcal{V}^2})^{-1}$, which has a non-trivial solution $B \neq 0$ for any $L > 0$. Setting $y = \sqrt{2B/m\hbar}/\mathcal{V}$ the compatibility condition reads $e^{2m\mathcal{V}Ly/\hbar^2} = 1/(1 - y^2)$. Using the values of L and $\mathcal{V}$ given in the text one obtains $e^{2m\mathcal{V}L/\hbar^2} \gg 1$, hence $2B\hbar^2/(m\mathcal{V}^2) \simeq 1$ within a good approximation, i.e. $B \simeq m\mathcal{V}^2/(2\hbar^2)$, which coincides with the result obtained in presence of a single thin well. This approximation is indeed equivalent to the limit of a large distance L (hence $e^{2m\mathcal{V}L/\hbar^2} \gg 1$) between the two Dirac delta functions; it can be easily verified that in the same limit one has $b \simeq 1$ and $a \simeq 0$, so that, in practice, the state is localized around the attractive delta function in $x = 0$, which is the binding part of the potential, and does not feel the presence of the other term in the potential which is very far away.

As L is decreased, the binding energy lowers and the wave function amplitude, hence the probability density, gets asymmetrically shifted on the left, until the binding energy vanishes in the limit $L \to 0$. In practice, the positive delta function in $x = L$ acts as a repulsive term which asymptotically extracts, as $L \to 0$, the particle from its thin well.

2.29. A particle of mass $M = 10^{-26}$ Kg moves along the x axis under the influence of an elastic force of constant $k = 10^{-6}$ N/m. The particle is in its fundamental state: compute its wave function and the mean value of x^2, given by

$$\langle x^2 \rangle = \frac{\int_{-\infty}^{\infty} dx\, x^2 |\psi(x)|^2}{\int_{-\infty}^{\infty} dx |\psi(x)|^2} \ .$$

Answer:

$$\psi(x) = \left(\frac{kM}{\pi^2\hbar^2}\right)^{1/8} e^{-\sqrt{kM}x^2/2\hbar} \ ; \qquad \langle x^2 \rangle = \frac{1}{2}\frac{\hbar}{\sqrt{kM}} \simeq 5\ 10^{-19} \text{m}^2$$

2.30. A particle of mass $M = 10^{-25}$ Kg moves in a 3-dimensional isotropic harmonic potential of elastic constant $k = 10$ N/m. How many states have energy less than $2\ 10^{-2}$ eV?

Answer: $E_{n_x \cdot n_y, n_z} = \hbar\sqrt{k/M}(3/2 + n_x + n_y + n_z)$. Therefore $E_{n_x \cdot n_y, n_z} < 2 \cdot 10^{-2}$ eV is equivalent to $n_x + n_y + n_z < 1.54$, corresponding to 4 possible states, $(n_x, n_y, n_z) = (0,0,0), (1,0,0), (0,1,0), (0,0,1)$.

2.31. A particle of mass $M = 10^{-26}$ Kg moves in one dimension under the influence of an elastic force of constant $k = 10^{-6}$ N/m and of a constant force $F = 10^{-15}$ N acting in the positive x direction. Compute the wave function of the fundamental state and the corresponding mean value of the coordinate x, given by

$$\langle x \rangle = \frac{\int_{-\infty}^{\infty} dx\, x |\psi(x)|^2}{\int_{-\infty}^{\infty} dx\, |\psi(x)|^2} \ .$$

Answer: As in the analogous classical case, the problem can be brought back to
a simple harmonic oscillator with the same mass and elastic constant by a simple
change of variable, $y = x - F/K$, which is equivalent to shifting the equilibrium
position of the oscillator. Hence the energy levels are the same of the harmonic
oscillator and the wave function of the fundamental state is

$$\psi(x) = \left(\frac{kM}{\pi^2 \hbar^2} \right)^{1/8} e^{-\frac{\sqrt{kM}}{2\hbar}(x - F/k)^2} \ ,$$

while $\langle x \rangle = F/k = 10^{-9}$ m.

2.32. A particle of mass $m = 10^{-30}$ Kg and kinetic energy equal to 50 eV hits
a square potential well of width $L = 2\ 10^{-10}$ m and depth $V = 1$ eV. What
is the reflection coefficient computed up to the first non-vanishing order in $\frac{V}{2E}$?

Answer: Let us suppose the square well to have endpoints $x = 0$ and $x = L$ and
that the potential vanishes outside the well. Let ψ_s, ψ_c and ψ_d be respectively the
wave functions for $x < 0$, $0 < x < L$ and $x > L$. If the particle comes from the left,
then $\psi_s = e^{i\sqrt{2mE}x/\hbar} + a\ e^{-i\sqrt{2mE}x/\hbar}$, $\psi_c = b\ e^{i\sqrt{2m(E+V)}x/\hbar} + c\ e^{-i\sqrt{2m(E+V)}x/\hbar}$
and $\psi_d = d\ e^{i\sqrt{2mE}x/\hbar}$ where a and c are necessarily of order $V/2E$ while b and d
are equal to 1 minus corrections of the same order. Indeed, as $V \to 0$ the solution
must tend to a single plane wave. By applying the continuity constraints we obtain:

$$1 + a = b + c \ ,$$

$$1 - a \simeq b - c + \frac{V}{2E} \ ,$$

$$be^{\frac{i\sqrt{2m(E+V)}L}{\hbar}} + ce^{-i\sqrt{2m(E+V)}L/\hbar} \simeq (b + \frac{V}{2E})e^{i\sqrt{2m(E+V)}L\hbar} - ce^{-i\sqrt{2m(E+V)}L/\hbar} \ ,$$

which are solved by $a \simeq \frac{V}{4E}(e^{2i\sqrt{2m(E+V)}L/\hbar} - 1)$ and

$$R = \frac{V^2}{4E^2} \sin^2 \frac{\sqrt{2m(E+V)}L}{\hbar} \simeq 0.96\ 10^{-4} \ .$$

2.33. A particle of mass $m = 10^{-30}$ Kg is confined inside a line segment of
length $L = 10^{-9}$ m with reflecting endpoints, which is centered around the
origin. In the middle of the line segment a thin repulsive potential barrier,
describable as $V(x) = W\delta(x)$, acts on the particle, with $W = 2\ 10^{-28}$ J m.
Compare the fundamental state of the particle with what found in absence of
the barrier.

Answer: Let us consider how the solutions of the Schrödinger equation in a line segment are influenced by the presence of the barrier. Odd solutions, contrary to even
ones, do not change since they vanish right in the middle of the segment, so that the
particle never feels the presence of the barrier. In order to discuss even solutions, let
us notice that they can be written, shifting the origin in the left end of the segment,

as $\psi_s \sim \sin\left(\sqrt{2mE}x/\hbar\right)$ for $x < L/2$ and $\psi_d \sim \sin\left(\sqrt{2mE}(L-x)/\hbar\right)$ for $x > L/2$. Setting $z \equiv \sqrt{2mEL}/(2\hbar)$ the discontinuity in the wave function derivative in the middle of the segment gives $\tan z = -z2\hbar^2/(mLW) \simeq -10^{-1}\,z$. Hence we obtain, for the fundamental state, $E \simeq \frac{2\hbar^2}{mL^2}\pi^2(1 - 2\ 10^{-1}) \simeq 2.75\ 10^{-19}$ J, slightly below the first excited level.

Notice that, increasing the intensity of the repulsive barrier W from 0 to ∞, the fundamental level grows from $\pi^2\hbar^2/(2mL^2)$ to $2\pi^2\hbar^2/(mL^2)$, i.e. it is degenerate with the first excited level in the $W \to \infty$ limit. There is no contradiction with the expected non-degeneracy since, in that limit, the barrier acts as a perfectly reflecting partition wall which separates the original line segment in two non-communicating segments: the two degenerate lowest states (as well as all the other excited ones) can thus be seen as two different superpositions (symmetric and antisymmetric) of the fundamental states of each segment.

2.34. An electron beam corresponding to an electric current $I = 10^{-12}$ A hits, coming from the right, the potential step sketched in the figure. The potential energy diverges for $x < 0$ while it is $-V = -10$ eV for $0 < x < L$ and 0 for $x > L$, with $L = 10^{-11}$ m . The kinetic energy kinetic energy of the electrons is $E_k = 0.01$ eV for $x > L$. Compute the electric charge density as a function of x.

Answer: There is complete reflection in $x = 0$, hence the current density is zero along the whole axis and we can consider a real wave function. In particular we set $\psi(x) = a\sin(\sqrt{2mE}(x-L)/\hbar + \phi)$ for $x > L$ and $\psi(x) = b\sin(\sqrt{2m(E+V)}x/\hbar)$ for $0 < x < L$. Continuity conditions read $b\sin(\sqrt{2m(E+V)}L/\hbar) = a\sin\phi \simeq b\sin(\sqrt{2mV}L/\hbar)$, $b\cos(\sqrt{2m(E+V)}L/\hbar) = \sqrt{E/(E+V)}a\cos\phi \simeq \sqrt{E/V}a\cos\phi \simeq b\cos(\sqrt{2mV}L/\hbar)$ (notice that $\sqrt{2mV}L/\hbar \simeq 0.57$ rad hence $\cos(\sqrt{2mV}L/\hbar) \simeq 0.85$). That fixes $\sqrt{E/V}\tan\left(\sqrt{2mV}L/\hbar\right) \simeq \tan\phi \simeq \phi$ and $b = a\sqrt{E/V}/\cos(\sqrt{2mV}L/\hbar)$, while the incident current fixes the value of a, $I = ea^2\sqrt{2E/m}$. Finally we can write, for the charge density,

$$e\rho = I\sqrt{\frac{m}{2E}}\sin^2\left(\frac{\sqrt{2mE}}{\hbar}(x-L) + \phi\right)$$

for $x > L$ and

$$e\rho \sim I\sqrt{\frac{mE}{2V^2}}\sin^2\left(\frac{\sqrt{2mE}}{\hbar}x\right)$$

for $0 < x < L$.

2.35. Referring to the potential energy given in Problem 2.34, determine the values of V for which there is one single bound state.

Answer: It can be easily realized that any possible bound states of the potential well considered in the problem will coincide with one of the odd bound states of the square well having the same depth and extending from $-L$ to L. The condition for the existence of a single bound state is therefore $\pi/2 < \sqrt{2mV}L/\hbar < 3\pi/2$.

2.36. A ball of mass $m = 0.05$ Kg moves at a speed of 3 m/s and without rolling towards a smooth barrier of thickness $D = 10$ cm and height $H = 1$ m. Using the formula for the tunnel effect, give a rough estimate about the probability of the ball getting through the barrier.

Answer: The transmission coefficient is roughly

$$T \sim \exp\left(-\frac{2D}{\hbar}\sqrt{2m(mgH - \frac{mv^2}{2})}\right) = 10^{-1.3 \ 10^{32}} .$$

2.37. What is the quantum of energy for a simple pendulum of length $l = 1$ m making small oscillations?

Answer: In the limit of small oscillations the pendulum can be described as a harmonic oscillator of frequency $\nu = 2\pi\omega = 2\pi\sqrt{g/l}$, where $g \simeq 9.8$ m/s is the gravitational acceleration on the Earth surface. The energy quantum is therefore $h\nu = 3.1 \ 10^{-34}$ J.

2.38. Compute the mean value of x^2 in the first excited state of a harmonic oscillator of elastic constant k and mass m.

Answer: The wave function of the first excited state is $\psi_1 \propto x \ e^{-x^2\sqrt{km/(4\hbar^2)}}$, hence

$$\langle x^2 \rangle = \frac{\int_{-\infty}^{\infty} x^4 e^{-x^2\sqrt{km/\hbar^2}} dx}{\int_{-\infty}^{\infty} x^2 e^{-x^2\sqrt{km/\hbar^2}} dx} = \frac{3}{2}\sqrt{\frac{\hbar^2}{km}} .$$

2.39. A particle of mass M moves in a line segment with reflecting endpoints placed at distance L. If the particle is in the first excited state $(n = 2)$, what is the mean quadratic deviation of the particle position from its average value, i.e. $\sqrt{\langle x^2 \rangle - \langle x \rangle^2}$?

Answer: Setting the origin in the middle of the segment, the wave function is $\psi(x) = \sqrt{2/L}\sin(2\pi x/L)$ inside the segment and vanishes outside. Obviously $\langle x \rangle = 0$ by symmetry, while

$$\langle x^2 \rangle = \frac{2}{L}\int_{-\frac{L}{2}}^{\frac{L}{2}} x^2 \sin^2\left(\frac{2\pi x}{L}\right) dx = L^2\left(\frac{1}{12} - \frac{1}{8\pi^2}\right)$$

whose square root gives the requested mean quadratic deviation.

2.40. An electron beam of energy E hits, coming from the left, the following potential barrier: $V(x) = \mathcal{V}\delta(x)$ where $\mathcal{V}$ is tuned to $\hbar\sqrt{2E/m}$. Compute the probability density on both sides of the barrier.

Answer: The wave function can be set to $e^{ikx} + a\, e^{-ikx}$ for $x < 0$ and to $b\, e^{ikx}$ for $x > 0$, where $k = \sqrt{2mE}/\hbar$. Continuity and discontinuity constraints for ψ and ψ' in $x = 0$ lead to

$$a = \frac{1}{\frac{ik\hbar^2}{mV} - 1} = \frac{1}{i - 1}, \qquad b = \frac{\frac{ik\hbar^2}{mV}}{\frac{ik\hbar^2}{mV} - 1} = \frac{1}{i + 1}.$$

The probability density is therefore $\rho = 1/2$ for $x > 0$, while for $x > 0$ it is $\rho = 3/2 - \sqrt{2}\sin(2kx + \pi/4)$.

2.41. A particle moves in one dimension under the influence of the potential given in Problem 2.34. Assuming that

$$\sqrt{2mV}\frac{L}{\hbar} = \frac{\pi}{2} + \delta, \quad \text{with} \quad \delta \ll 1,$$

show that, at the first non-vanishing order in δ, one has $B \simeq V\delta^2$, where $B = -E$ and E is the energy of the bound state. Compute the ratio of the probability of the particle being inside the well to that of being outside.

Answer: The depth of the potential is slightly above the minimum for having at least one bound state (see the solution of Problem 2.36), therefore we expect a small binding energy. In particular the equation for the bound state energy, which can be derived by imposing the continuity constraints, is $\cot\left(\sqrt{2m(V - B)}L/\hbar\right) = -\sqrt{B/(V - B)}$. The particular choice of parameters implies $B \ll V$, hence $\cot\left(\sqrt{2m(V - B)}L/\hbar\right) \simeq \cot(\pi/2(1 - B/(2V)) + \delta) \simeq -\delta + \pi B/(4V) \simeq -\sqrt{B/V}$ hence $B \simeq V\delta^2$. Therefore the wave function is well approximated by $k \sin \pi x/2L$ inside the well and by $ke^{-\sqrt{2mB}(x-L)/\hbar} \simeq ke^{-\pi\delta(x-L)/2L}$ outside, where k is a normalization constant. The ratio of probabilities is $\pi\delta/2$: the very small binding energy is reflected in the large probability of finding the particle outside the well.

3

Introduction to the Statistical Theory of Matter

In Chapter 2 we have discussed the existence and the order of magnitude of quantum effects, showing in particular their importance for microscopic physics. We have seen that quantum effects are relevant for electrons at energies of the order of the electron-Volt, while for the dynamics of atoms in crystals, which have masses three or four order of magnitudes larger, significant effects appear at considerably lower energies, corresponding to low temperatures. Hence, in order to study these effects, a proper theoretical framework is needed for describing systems made up of particles at thermal equilibrium and for deducing their thermodynamical properties from the (quantum) nature of their states.

Boltzmann identified the thermal contact among systems as a series of shortly lasting and random interactions with limited energy exchange. These interactions can be assimilated to collisions among components of two different systems taking place at the surface of the systems themselves. Collisions generate sudden transitions among the possible states of motion for the parts involved. The sequence of collision processes is therefore analogous to a series of dice casts by which subsequent states of motion are chosen by drawing lots.

It is clear that in these conditions it is not sensible to study the time evolution of the system, since that is nothing but a random succession of states of motion. Instead it makes sense to study the *distribution* of states among those accessible to each system, i.e. the number of times a particular state occurs in **N** different observations. In case of completely random transitions among all the possible states, the above number is independent of the particular state considered and equal to the number of observations **N** divided by the total number of possible states. In place of the distribution of results of subsequent observations we can think of the *distribution of the probability* that the system be in a given state: under the same hypothesis of complete randomness, the probability distribution is independent of the state and equal to the inverse of the total number of accessible states.

However the problem is more complicated if we try to take energy conservation into account. Although the energy exchanged in a typical microscopic collision is very small, the global amount of energy (heat) transferred in a great number of interactions can be macroscopically relevant; on the other hand, the total energy of all interacting macroscopic systems must be constant.

The american physicist J. Willard Gibbs proposed a method to evaluate the probability distribution for the various possible states in the case of thermal equilibrium[1]. His method is based on the following points:

1) Thermal equilibrium is independent of the nature of the heat reservoir, which must be identified with a system having infinite thermal capacity (that is a possible enunciation of the so-called *zero-th principle of thermodynamics*). According to Gibbs, the heat reservoir is a set of $\mathcal{N}$ systems, identical to the one under consideration, which are placed in thermal contact. $\mathcal{N}$ is so great that each heat (energy) exchange between the system and the reservoir, being distributed among all different constituent systems, does not alter their average energy content, hence their thermodynamical state.

2) Gibbs assumed transitions to be induced by completely random collisions. Instead of following the result of a long series of random transitions among states, thus extracting the probability distribution by averaging over time evolution histories, Gibbs proposed to consider a large number of simultaneous draws and to take the average over them. That is analogous to drawing a large number of dice simultaneously instead of a single die for a large number of times: time averages are substituted by *ensemble averages*. Since in Gibbs scheme the system–reservoir pair (*Macrosystem*) can be identified with the $\mathcal{N} + 1$ identical systems in thermal contact and in equilibrium, if at any time the distribution of the states occupied by the various systems is measured, one has automatically an average over the ensemble and the occupation probability for the possible states of a single system can be deduced. On the other hand, computing the ensemble distribution does not require the knowledge of the state of each single system, but instead that of the number of systems in each possible state.

3) The macrosystem is isolated and internal collisions induce random changes of its state. However, all possible macrosystem states with the same total energy are assumed to be equally probable. That clearly implies that the probability associated to a given distribution of the $\mathcal{N} + 1$ systems is directly proportional to the number of states of the macrosystem realizing the given distribution: this number is usually called *multiplicity*. If i is the index distinguishing all possible system states, any distribution is fixed by a succession of integers $\{N_i\}$, where N_i is the number of systems occupying state i. It is easy to verify that the multiplicity $\mathcal{M}$ is given by

---

[1] Notice that the states considered by Gibbs in the XIX century were small cells in the space of states of motion (the phase space) of the system, while we shall consider quantum states corresponding to independent solutions of the stationary Schrödinger equation for the system. This roughly corresponds to choosing the volume of Gibbs cells of the order of magnitude of $h^{\mathcal{N}}$, where $\mathcal{N}$ is the number of degrees of freedom of the system.

$$\mathcal{M}(\{N_i\}) = \frac{\mathcal{N}!}{\prod_i N_i!}, \tag{3.1}$$

with the obvious constraint

$$\sum_i N_i = \mathcal{N}. \tag{3.2}$$

4) The accessibility criterion for states is solely related to their energy which, due to the limited energy exchange in collision processes, is reduced to the sum of the energies of the constituent systems. Stated otherwise, if E_i is the energy of a single constituent system when it is in state i, disregarding the interaction energy between systems, the total energy of the macrosystem is identified with:

$$E_{\text{tot}} = \sum_i E_i N_i \equiv \mathcal{N}U. \tag{3.3}$$

Thus U can be identified with the average energy of the constituent systems: it characterizes the thermodynamical state of the reservoir and must therefore be related to its temperature in some way to be determined by computations.

5) Gibbs identified the probability of the considered system being in state i with:

$$p_i = \frac{\bar{N}_i}{\mathcal{N}}, \tag{3.4}$$

where the distribution $\{\bar{N}_i\}$ is the one having maximum multiplicity among all possible distributions:

$$\mathcal{M}(\{\bar{N}_i\}) \geq \mathcal{M}(\{N_i\}) \quad \forall \ \{N_i\},$$

i.e. that is realized by the largest number of states of the macrosystem. We call p_i the *occupation probability* occupation of state i.

The identification made by Gibbs is justified by the fact that the multiplicity function has only one sharp peak in correspondence of its maximum whose width $(\Delta\mathcal{M}/\mathcal{M})$ vanishes in the limit of an ideal reservoir, i.e. as $\mathcal{N} \to \infty$. Later on we shall discuss a very simple example, even if not very significant from the physical point of view, corresponding to a system with only three possible states, so that the multiplicity $\mathcal{M}$, given the two constraints in (3.2) and (3.3), will be a function of a single variable, thus allowing an easy computation of the width of the peak.

6) The analysis of thermodynamical equilibrium described above can be extended to the case in which also the number of particles in each system is variable: not only energy transfer by collisions can take place at the surfaces of the systems, but also exchange of particles of various species (atoms, molecules, electrons, ions and so on). In this case the various possible states of the system are characterized not only by their energy but also by the number of particles of each considered species. We will indicate by $n_i^{(s)}$ the number of particles belonging to species s present in state i; therefore, besides the energy E_i, there will be as many fixed quantities as the number of possible species

characterizing each possible state of the system. The distribution of states in the macrosystem, $\{N_i\}$, will then be subject to further constraints, besides (3.2) and (3.3), related to the conservation of the total number of particles for each species, namely

$$\sum_i n_i^{(s)} N_i = \bar{n}^{(s)} \mathcal{N} \tag{3.5}$$

for each s. The kind of thermodynamical equilibrium described in this case is very different from the previous one. While in the first case equilibrium corresponds to the system and reservoir having the same temperature ($T_1 = T_2$), in the second case also Gibbs potential $g^{(s)}$ will be equal, for each constrained single particle species separately. In place of $g^{(s)}$ it is usual to consider the quantity known as *chemical potential*, defined as $\mu^{(s)} \equiv g^{(s)}/N_A$, where $N_A = 6.02 \; 10^{23}$ is Avogadro's number.

The distribution corresponding to the two different kind of equilibrium are named differently. For a purely thermal equilibrium we speak of the *Canonical Distribution*, while when considering also particle number equilibrium we speak of the *Grand Canonical Distribution*. We will start by studying simple systems by means of the Canonical Distribution and will then make use of the Grand Canonical Distribution for the case of perfect quantum gasses.

As the simplest possible systems we shall consider in particular an isotropic three-dimensional harmonic oscillator (*Einstein's crystal*) and a particle confined in a box with reflecting walls. Let us briefly recall the nature of the states for the two systems.

Einstein's crystal

In this model atoms do not exchange forces among themselves but in rare collisions whose nature is not well specified and whose only role is that of assuring thermal equilibrium. Atoms are instead attracted by elastic forces towards fixed points corresponding to the vertices of a crystal lattice.

The attraction point for the generic atom is identified by the coordinates $(m_x a, m_y a, m_z a)$, where m_x, m_y, m_z are relative integer numbers with $|m_i| a < L/2$: L is the linear size and a is the spacing of the crystal lattice, which is assumed to be cubic. To summarize, each atom corresponds to a vector $\boldsymbol{m}$ of components m_x, m_y, m_z.

Hence Einstein's crystal is equivalent to a large number of isotropic harmonic oscillators and can be identified with the macrosystem itself. According to the analysis of the harmonic oscillator made in previous Chapter, the *microscopic* quantum state of the crystal is characterized by three non-negative integer numbers $(n_{x,\boldsymbol{m}}, n_{y,\boldsymbol{m}}, n_{z,\boldsymbol{m}})$ for every vertex ($\boldsymbol{m}$). The corresponding energy level is given by

$$E_{n_{x,\boldsymbol{m}}, n_{y,\boldsymbol{m}}, n_{z,\boldsymbol{m}}} = \sum_{\boldsymbol{m}} \hbar \omega \left(n_{x,\boldsymbol{m}} + n_{y,\boldsymbol{m}} + n_{z,\boldsymbol{m}} + \frac{3}{2} \right). \tag{3.6}$$

It is clear that several different states correspond to the same energy level: following the same notation as in Chapter 2, they are called *degenerate*.

In our analysis of the harmonic oscillator we have seen that states described as above correspond to solutions of the stationary Schrödinger equation, i.e. to wave functions depending on time through the phase factor $e^{-iEt/\hbar}$. Therefore the state of the macrosystem would not change in absence of further interactions among the various oscillators, and the statistical analysis would make no sense. If we instead admit the existence of rare random collisions among the oscillators leading to small energy exchanges, then the state of the macrosystem evolves while its total energy stays constant.

The particle in a box with reflecting walls

In this case the reservoir is made up of $\mathcal{N}$ different boxes, each containing one particle. Energy is transferred from one box to another by an unspecified collisional mechanism acting through the walls of the boxes. We have seen in previous Chapter that the quantum states of a particle in a box are described by three positive integer numbers (k_x, k_y, k_z), which are related to the wave number components of the particle and correspond to an energy

$$E_{\mathbf{k}} = \frac{\hbar^2 \pi^2}{2mL^2} [k_x^2 + k_y^2 + k_z^2] \ . \tag{3.7}$$

3.1 Thermal Equilibrium by Gibbs' Method

Following Gibbs' description given above, let us consider a system whose states are enumerated by an index i and have energy E_i. We are interested in the distribution which maximizes the multiplicity $\mathcal{M}$ defined in (3.1) when the constraints in (3.2) and (3.3) are taken into account. Since $\mathcal{M}$ is always positive, in place of it we can maximize its logarithm

$$\ln \mathcal{M}(\{N_i\}) = \ln \mathcal{N}! - \sum_i \ln N_i! \ . \tag{3.8}$$

If $\mathcal{N}$ is very large, thus approaching the so-called *Thermodynamical Limit* corresponding to an ideal reservoir, and if distributions corresponding to negligible multiplicity are excluded, we can assume that all N_i's get large as well. In these conditions we are allowed to replace factorials by Stirling formula:

$$\ln N! \simeq N \left(\ln N - 1 \right) \ . \tag{3.9}$$

If we set $N_i \equiv \mathcal{N} x_i$, then the logarithm of the multiplicity is approximately

$$\ln \mathcal{M}(\{N_i\}) \simeq -\mathcal{N} \sum_i x_i (\ln x_i - 1) \ , \tag{3.10}$$

and the constraints in (3.2) and (3.3) are rewritten as:

$$\sum_i x_i = 1\,,$$

$$\sum_i E_i x_i = U\,. \tag{3.11}$$

In order to look for the maximum of expression (3.10) in presence of the constraints (3.11), it is convenient to apply the method of Lagrange's multipliers.

Let us remind that the stationary points of the function $F(x_1, .., x_n)$ in presence of the constraints: $G_j(x_1, .., x_n) = 0$, with $j = 1, .., k$ and $k < n$, are solutions of the following system of equations:

$$\frac{\partial}{\partial x_i} \left[F(x_1, .., x_n) + \sum_{j=1}^{k} \lambda_j G_j(x_1, .., x_n) \right] = 0\,, \quad i = 1, ..., n\,,$$

and of course of the constraints themselves. Therefore we have $n + k$ equations in $n + k$ variables x_i, $i = 1, ...n$, and λ_j, $j = 1, .., k$. In the generic case, both the unknown variables x_i and the multipliers λ_j will be determined univocally. In our case the system reads:

$$\frac{\partial}{\partial x_i} \left[\ln \mathcal{M} - \beta \mathcal{N} (\sum_j E_j x_j - U) + \alpha \mathcal{N} (\sum_j x_j - 1) \right]$$

$$= -\mathcal{N} \frac{\partial}{\partial x_i} \sum_j [x_j (\ln x_j - 1) + \beta E_j x_j - \alpha x_j]$$

$$= -\mathcal{N} [\ln x_i + \beta E_i - \alpha] = 0 \tag{3.12}$$

where $-\beta$ and α are the Lagrange multiplierss which can be computed by making use of (3.11).

Taking into account (3.4) and the discussion given in the introduction to this Chapter, we can identify the variables x_i solving system (3.12) with the occupation probabilities p_i in the Canonical Distribution, thus obtaining

$$\ln p_i + 1 + \beta E_i - \alpha = 0\,, \tag{3.13}$$

hence

$$p_i = \mathrm{e}^{-1 - \beta E_i + \alpha} \equiv k\,\mathrm{e}^{-\beta E_i}\,, \tag{3.14}$$

where β must necessarily be positive in order that the sums in (3.11) be convergent. The constraints give:

$$p_i = \frac{\mathrm{e}^{-\beta E_i}}{\sum_j \mathrm{e}^{-\beta E_j}} = -\frac{1}{\beta} \frac{d}{dE_i} \ln \sum_j \mathrm{e}^{-\beta E_j} \equiv -\frac{1}{\beta} \frac{d}{dE_i} \ln Z\,,$$

$$U = \sum_i E_i p_i = -\frac{d}{d\beta} \ln \sum_j \mathrm{e}^{-\beta E_j} \equiv -\frac{d}{d\beta} \ln Z\,, \tag{3.15}$$

where we have introduced the function

$$Z \equiv \sum_j e^{-\beta E_j} ,\tag{3.16}$$

which is known as the *partition function*.

The second of equations (3.15) expresses the relation between the Lagrange multiplier β and the average energy U, hence implicitly between β and the equilibrium temperature. It can be easily realized that in fact β is a universal function of the temperature, which is independent of the particular system under consideration.

To show that, let us consider the case in which each component system S can be actually described in terms of two independent systems s and s', whose possible states are indicated by the indices i and a corresponding to energies e_i and ϵ_a. The states of S are therefore described by the pair (a, i) corresponding to the energy:

$$E_{a,i} = \epsilon_a + e_i .$$

If we give the distribution in terms of the variables $x_{a,i} \equiv N_{a,i}/\mathcal{N}$ and we repeat previous analysis, we end up with searching for the maximum of

$$\ln \mathcal{M}(\{N_{a,i}\}) \simeq -\mathcal{N} \sum_{b,j} x_{b,j}(\ln x_{b,j} - 1) ,\tag{3.17}$$

constrained by:

$$\sum_{b,j} x_{b,j} = 1 ,$$

$$\sum_{b,j} (\epsilon_b + e_j) x_{b,j} = U .\tag{3.18}$$

Following previous analysis we finally find:

$$p_{a,i} = -\frac{1}{\beta} \frac{d}{dE_{a,i}} \ln Z ,$$

$$U = -\frac{d}{d\beta} \ln Z ,\tag{3.19}$$

but Z is now given by:

$$Z = \sum_{b,j} e^{-\beta(\epsilon_b + e_j)} = \sum_b \sum_j e^{-\beta \epsilon_b} e^{-\beta e_j} = \sum_b e^{-\beta \epsilon_b} \sum_j e^{-\beta e_j} = Z_s Z_{s'}\tag{3.20}$$

so that the occupation probability factorizes as follows:

$$p_{a,i} = \frac{e^{-\beta(\epsilon_a + e_i)}}{Z} = \frac{e^{-\beta \epsilon_a}}{Z_s} \frac{e^{-\beta e_i}}{Z_{s'}} = p_a \, p_i .$$

We have therefore learned that the two systems have independent distributions, but corresponding to the same value of β: that is a direct consequence of having written a single constraint on the total energy (leading to a single Lagrange multiplier β linked to energy conservation), and on its turn this is an implicit way of stating that the two systems are in contact with the same heat reservoir, i.e. that they have the same temperature: hence we conclude that β is a universal function of the temperature, $\beta = \beta(T)$. We shall explicitly exploit the fact that $\beta(T)$ is independent of the system under consideration by finding its exact form through the application of Gibbs method to systems as simple as possible.

3.1.1 Einstein's Crystal

Let us consider a little cube with an edge of length L and, following Gibbs method, let us put it in thermal contact with a great (infinite) number of similar little cubes, thus building an infinite crystal corresponding to the macrosystem, which is therefore imagined as divided into many little cubes. Actually, since by hypothesis single atoms do not interact with each other but in very rare thermalizing collisions, we can consider the little cube so small as to contain a single atom, which is then identified with an isotropic harmonic oscillator: we will obtain the occupation probability of its microscopic states at equilibrium and evidently the properties of a larger cube can be deduced by combining those of the single atoms independently.

We recall that the microscopic states of the oscillator are associated with a vector $\mathbf{n}$ having integer non-negative components, the corresponding energy level being given in (2.123). We can then easily compute the partition function of the single oscillator:

$$Z_o = \sum_{n_x=0}^{\infty} \sum_{n_y=0}^{\infty} \sum_{n_z=0}^{\infty} e^{-\beta\hbar\omega(n_x+n_y+n_z+3/2)} = e^{-3\beta\hbar\omega/2} \left[\sum_{n=0}^{\infty} \left(e^{-\beta\hbar\omega} \right)^n \right]^3$$

$$= \left[\frac{e^{-\beta\hbar\omega/2}}{1 - e^{-\beta\hbar\omega}} \right]^3 = \left[\frac{e^{\beta\hbar\omega/2}}{e^{\beta\hbar\omega} - 1} \right]^3 . \tag{3.21}$$

The average energy is then:

$$U = -\frac{\partial}{\partial\beta} \ln Z_o = -3\frac{\partial}{\partial\beta} \ln \frac{e^{\beta\hbar\omega/2}}{e^{\beta\hbar\omega} - 1} = \frac{3\hbar\omega}{2} \left(\frac{e^{\beta\hbar\omega} + 1}{e^{\beta\hbar\omega} - 1} \right) . \tag{3.22}$$

Approaching the classical limit, in which $\hbar \to 0$, this result gives direct information on β. Indeed in the classical limit Dulong–Petit's law must hold, stating that $U = 3kT$ where k is Boltzmann's constant. Instead, since $e^{\beta\hbar\omega} \simeq 1 + \beta\hbar\omega$ as $\hbar \to 0$, our formula tells us that in the classical limit we have $U \simeq 3/\beta$, which is also the result we would have obtained by directly applying Gibbs method to a classical harmonic oscillator (see Problem 3.11). Therefore we must identify

$$\beta = \frac{1}{kT}.$$

This result will be confirmed later by studying the statistical thermodynamics of perfect gasses. The specific heat, defined as

$$C = \frac{\partial U}{\partial T},$$

can then be computed through (3.22):

$$C = \frac{\partial \beta}{\partial T}\frac{\partial U}{\partial \beta} = -\frac{1}{kT^2}\frac{3}{2}\hbar^2\omega^2 e^{\beta\hbar\omega}\left[\frac{1}{e^{\beta\hbar\omega}-1} - \frac{e^{\beta\hbar\omega}+1}{\left(e^{\beta\hbar\omega}-1\right)^2}\right]$$

$$= \frac{3\hbar^2\omega^2}{kT^2}\frac{e^{\hbar\omega/kT}}{\left(e^{\hbar\omega/kT}-1\right)^2}. \tag{3.23}$$

Setting $x = kT/\hbar\omega$, the behavior of the atomic specific heat is shown in Fig. 3.1. It is clearly visible that when $x \geq 1$ Dulong–Petit's law is reproduced within a very good approximation. The importance of Einstein's model consists in having given the first qualitative explanation of the violations of the same law at low temperatures, in agreement with experimental measurements showing atomic specific heats systematically below $3k$. Einstein was the first showing that the specific heat vanishes at low temperatures, even if failing in predicting the exact behavior: that is cubic in T in insulators and linear in conductors (if the superconductor transition is not taken into account), while Einstein's model predicts C vanishing like $e^{-\hbar\omega/kT}$. The different behaviour can be explained, in the insulator case, through the fact that the hypothesis of single atoms being independent of each other, which is at the basis of Einstein's model, is far from being realistic. As a matter of fact, atoms move under the influence of forces exerted by nearby atoms, so that the crystal lattice itself is elastic and not rigid, as assumed in the model. In the case of conductors, instead, the linear behaviour is due to electrons in the conducting band.

Notwithstanding the wrong quantitative prediction, Einstein's model furnishes the correct qualitative interpretation for the vanishing of the specific heat as $T \to 0$. Since typical thermal energy exchanges are of the order of kT and since the system can only exchange quanta of energy equal to $\hbar\omega$, we infer that if $kT \ll \hbar\omega$ then quantum effects suppress the energy exchange between the system and the reservoir: the system cannot absorb any energy as the temperature is increased starting from zero, hence its specific heat vanishes. Notice also that quantum effects disappear as $kT \gg \hbar\omega$, when the typical energy exchange is much larger then the energy level spacing of the harmonic oscillator: energy quantization is not visible any more and the system behaves as if it were a classical oscillator. It is therefore from the comparison between the quantum energy scale of the system and the thermal energy scale, hence from the parameter $\hbar\omega/kT$, that we can infer if quantum effects are important or not.

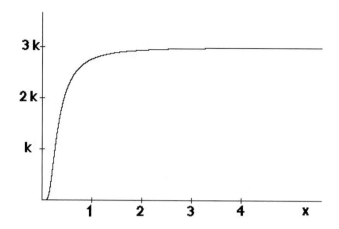

Fig. 3.1. A plot of the atomic specific heat in Einstein's model in k units as a function of $x = \hbar\omega/kT$, showing the vanishing of the specific heat at low temperatures and its asymptotic agreement with Dulong–Petit's value $3k$

3.1.2 The Particle in a Box with Reflecting Walls

In this case Gibbs reservoir is made up of $\mathcal{N}$ boxes of size L. The state of the particle in the box is specified by a vector with positive integer components k_x, k_y, k_z corresponding to the energy given in (3.7). The partition function of the system is therefore given by

$$Z = \sum_{k_x=1}^{\infty} \sum_{k_y=1}^{\infty} \sum_{k_z=1}^{\infty} e^{-\beta \frac{\hbar^2 \pi^2}{2mL^2}(k_x^2+k_y^2+k_z^2)} = \left(\sum_{k=1}^{\infty} e^{-\beta \frac{\hbar^2 \pi^2}{2mL^2} k^2} \right)^3. \qquad (3.24)$$

For large values of β, hence for small temperatures, we have:

$$Z \simeq \left(e^{-\beta \frac{\hbar^2 \pi^2}{2mL^2}} \right)^3, \qquad (3.25)$$

because the first term in the series on the right hand side of (3.24) dominates over the others. Instead for large temperatures, noticing that the quantity which is summed up in (3.24) changes very slowly as a function of $\boldsymbol{k}$, we can replace the sum by an integral:

$$Z = \left(\sum_{k=1}^{\infty} e^{-\beta \frac{\hbar^2 \pi^2}{2mL^2} k^2} \right)^3 \simeq \left(\int_0^{\infty} dk \; e^{-\beta \frac{\hbar^2 \pi^2}{2mL^2} k^2} \right)^3$$

$$= \left(\frac{2mL^2}{\beta \pi^2 \hbar^2} \right)^{\frac{3}{2}} \left(\int_0^{\infty} dx e^{-x^2} \right)^3 = \left(\frac{m}{2\beta\pi} \right)^{\frac{3}{2}} \frac{L^3}{\hbar^3} = \left(\frac{m}{2\beta\pi\hbar^2} \right)^{\frac{3}{2}} V . \quad (3.26)$$

We conclude that while at low temperatures $(\beta \frac{\hbar^2 \pi^2}{2mL^2} \gg 1)$ the mean energy tends to

$$U \to -\frac{d}{d\beta}\left(-3\beta \frac{\hbar^2 \pi^2}{2mL^2}\right) = 3\frac{\hbar^2 \pi^2}{2mL^2}, \tag{3.27}$$

i.e. to the energy of the fundamental state of the system, at high temperatures $(\beta \frac{\hbar^2 \pi^2}{2mL^2} \ll 1)$ we have

$$U \to -\frac{d}{d\beta}\left[\ln V + \frac{3}{2}\left(\ln m - \ln \beta - \ln(2\pi\hbar^2)\right)\right] = \frac{3}{2\beta} = \frac{3}{2}kT. \tag{3.28}$$

This confirms that $\beta = \frac{1}{kT}$, since the system under consideration corresponds to a perfect gas containing a single atom, whose mean energy in the classical limit is precisely $3kT/2$, if T is the absolute temperature.

3.2 The Pressure and the Equation of State

It is well known that the equation of state for a homogeneous and isotropic system fixes a relation among the pressure, the volume and the temperature of the same system when it is at thermal equilibrium. We will discuss now how the pressure of the system can be computed once the distribution over its states is known.

The starting point for the computation of the pressure is a theorem which is valid both in classical and quantum mechanics and is known as the *adiabatic theorem* . In the quantum version the theorem makes reference to a system defined by parameters which change very slowly in time (where it is understood that a change in the parameters may also change the energy levels of the system) and asserts that in these conditions the system maintains its quantum numbers unchanged. What is still unspecified in the enunciation above is what is the meaning of *slow*, i.e. with respect to what time scale. We shall clarify that by an example.

Let us consider a particle of mass M moving in a one-dimensional segment of length L with reflecting endpoints. Suppose the particle is in the n-th quantum state corresponding to the energy $E_n = \hbar^2 n^2 \pi^2/(2ML^2)$ (see equation (2.97)) and that we slowly reduce the distance L, where slowly means that $|\delta L|/L \ll 1$ in a time interval of the order of $\hbar/\Delta E_n$, where ΔE_n is the energy difference between two successive levels. In this case the adiabatic theorem applies and states that the particle keeps staying in the n-th level as L is changed. That of course means that the energy of the particle increases as we bring the two endpoints closer to each other, $\delta E_n = (\partial E_n/\partial L)\delta L$, and we can interpret this energy variation as the work that we must do to move them. On the other hand, assuming the endpoints to be practically massless, in order to move them slowly we must exactly balance the force exerted on them by the presence of the particle inside, which is the analogous of the pressure in the one-dimensional case. Therefore we obtain the following force:

$$F(n, L) = -\frac{dE_n(L)}{dL} = \frac{\hbar^2 n^2 \pi^2}{ML^3} . \tag{3.29}$$

If we now consider the three-dimensional case of the particle in a box of volume $V = L^3$, for which, according to (2.101), we have:

$$E_n(V) = \frac{\hbar^2 |n|^2 \pi^2}{2ML^2} = \frac{\hbar^2 |n|^2 \pi^2}{2MV^{\frac{2}{3}}} , \tag{3.30}$$

we can generalize (3.29) replacing the force by the pressure:

$$P(k, V) = -\frac{dE_k(V)}{dV} = -\frac{1}{3L^2}\frac{dE_k}{dL} = \frac{2}{3}\frac{E_k(V)}{V} . \tag{3.31}$$

Our choice to consider the pressure P instead of the force is dictated by our intention of treating the system without making explicit reference to the specific orientation of the cubic box. The force is equally exerted on all the walls of the box and is proportional to the area of each box, the pressure being the coefficient of proportionality.

Having learned how to compute the pressure when the system is in one particular quantum state, we notice that at thermal equilibrium, being the i-th state occupied with the probability p_i given in (3.15), the pressure can be computed by averaging that obtained for the single state over the Canonical Distribution, thus obtaining:

$$P = -\frac{1}{Z}\sum_i e^{-\beta E_i(V)}\frac{\partial E_i}{\partial V} . \tag{3.32}$$

In the case of a particle in a cubic box, using the result of (3.31), we can write:

$$P = \frac{2}{3V}\sum_{k_x,k_y,k_z=0}^{\infty}\frac{e^{-\beta E_k(V)}}{Z}E_k = \frac{2U}{3V} , \tag{3.33}$$

which represents the equation of state of our system. In the high temperature limit, taking into account (3.28), we easily obtain

$$P = \frac{kT}{V} , \tag{3.34}$$

which coincides with the equation of state of a classical perfect gas made up of a single atom in a volume V.

Starting from the definition of the partition function Z in (3.16), we can translate (3.32) into a formula of general validity:

$$P = \frac{1}{\beta}\frac{\partial \ln Z}{\partial V} . \tag{3.35}$$

3.3 A Three Level System

In order to further illustrate Gibbs method and in particular to verify what already stated about the behavior of the multiplicity function in the limit of an ideal reservoir, $\mathcal{N} \to \infty$ (i.e. that it has only one sharp peak in correspondence of its maximum whose width vanishes in that limit), let us consider a very simple system characterized by three energy levels $E_1 = 0$, $E_2 = \epsilon$ and $E_3 = 2\epsilon$, each corresponding to a single microscopic state. Let $\mathcal{U}$ be the total energy of the macrosystem containing $\mathcal{N}$ copies of the system; it is obvious that $\mathcal{U} \leq 2\mathcal{N}\epsilon$. The statistical distribution is fixed by giving the number of copies in each microscopic state, i.e. by three non-negative integer numbers N_1, N_2, N_3 constrained by:

$$N_1 + N_2 + N_3 = \mathcal{N}, \qquad (3.36)$$

and by:

$$N_2\epsilon + 2N_3\epsilon = \mathcal{U}. \qquad (3.37)$$

the multiplicity of the distribution is

$$\mathcal{M}_{(N_i)} \equiv \frac{\mathcal{N}!}{N_1!N_2!N_3!}. \qquad (3.38)$$

The simplicity of the model lies in the fact that, due to the constraints, there is actually only one free variable among N_1, N_2 and N_3 on which the distribution is dependent. In particular we choose N_3 and parametrize it as

$$N_3 = x\mathcal{N}.$$

Solving the constraints given above we have:

$$N_2 = \frac{\mathcal{U}}{\epsilon} - 2x\mathcal{N} \equiv (u - 2x)\mathcal{N},$$

where the quantity $u \equiv \mathcal{U}/(\epsilon\mathcal{N})$ has been introduced, which is proportional to the mean energy of the copies ($U = u\epsilon$), and:

$$N_1 = (1 - u + x)\mathcal{N}.$$

The fact that the occupation numbers N_i are non-negative integers implies that x must be greater than the maximum between 0 and $u - 1$, and less than $u/2$. Notice also that if $u > 1$ then $N_3 > N_1$. We must exclude this possibility since, as it is clear from the expression of the Canonical Distribution in (3.15), at thermodynamical equilibrium occupation numbers must decrease as the energy increases. There are however cases, like for instance those encountered in *laser* physics, in which the distributions are really reversed (i.e. the most populated levels are those having the highest energies), but they correspond to situations which are not at thermal equilibrium.

In the thermodynamical limit $\mathcal{N} \to \infty$, we can also say, neglecting distributions with small multiplicities, that each occupation number N_i becomes very large, so that we can rewrite factorials by using Stirling's formula, equation (3.9), and the expression giving the multiplicity as:

$$\mathcal{M}(x) \simeq c \frac{\mathcal{N}^{\mathcal{N}}}{(x\mathcal{N})^{x\mathcal{N}} ((u - 2x)\mathcal{N})^{(u-2x)\mathcal{N}} ((1 - u + x)\mathcal{N})^{(1-u+x)\mathcal{N}}}$$

$$= c \left(x^{-x} (u - 2x)^{-(u-2x)} (1 - u + x)^{-(1-u+x)} \right)^{\mathcal{N}}, \qquad (3.39)$$

where c is a constant, which will not enter our considerations.

The important point in our analysis is that the expression in brackets in (3.39) is positive in the allowed range $0 \leq x \leq u/2$ and has a single maximum, which is strictly inside that range. To find its position we can therefore study, in place of $\mathcal{M}$, its logarithm

$$\ln \mathcal{M}(x) \simeq -\mathcal{N} \left(x \ln x + (u - 2x) \ln(u - 2x) + (1 - u + x) \ln(1 - u + x) \right),$$

whose derivative is

$$(\ln \mathcal{M}(x))' = -\mathcal{N} \left(\ln x - 2 \ln(u - 2x) + \ln(1 - u + x) \right),$$

and vanishes if

$$x(1 - u + x) = (u - 2x)^2. \qquad (3.40)$$

Equation (3.40) shows that, in the most likely distribution, N_2 is the geometric mean of N_1 and N_3. Hence there is a number $z < 1$ such that $N_2 = z N_1$ and $N_3 = z^2 N_1$. According to the Canonical Distribution, $z = e^{-\beta \epsilon}$. Equation (3.40) has real solutions:

$$x = \frac{1 + 3u \pm \sqrt{(1 + 3u)^2 - 12u^2}}{6}.$$

The one contained in the allowed range $0 \leq x \leq u/2$ is

$$x_M = \frac{1 + 3u - \sqrt{(1 + 3u)^2 - 12u^2}}{6}.$$

We can compute the second derivative of the multiplicity in x_M by using the relation:

$$\mathcal{M}'(x) = -\mathcal{N} \left(\ln x - 2 \ln(u - 2x) + \ln(1 - u + x) \right) \mathcal{M}(x),$$

and, obviously, $\mathcal{M}'(x_M) = 0$. Taking into account (3.40) we have then:

$$\mathcal{M}''(x_M) = -\mathcal{N} \left(\frac{1}{x_M} + \frac{4}{u - 2x_M} + \frac{1}{1 + x_M - u} \right) \mathcal{M}(x_M)$$

$$= -\mathcal{N} \frac{1 + 3u - 6x_M}{(u - 2x_M)^2} \mathcal{M}(x_M). \qquad (3.41)$$

Replacing the value found for x_M we arrive finally to:

$$\frac{\mathcal{M}''(x_M)}{\mathcal{M}(x_M)} = -\mathcal{N}\frac{9\sqrt{(1+3u)^2 - 12u^2}}{\left(\sqrt{(1+3u)^2 - 12u^2} - 1\right)^2} \leq -18\mathcal{N}. \tag{3.42}$$

This result, and in particular the fact that $\mathcal{M}''(x_M)$ is of the order of $-\mathcal{N}\mathcal{M}(x_M)$ as $\mathcal{N} \to \infty$, so that

$$\frac{\mathcal{M}(x_M) - \mathcal{M}(x)}{\mathcal{M}(x_M)} \sim \mathcal{N}(x - x_M)^2,$$

demonstrates that the multiplicity has a maximum whose width goes to zero as $1/\sqrt{\mathcal{N}}$. That is also clearly illustrated in Fig. 3.2. Therefore, in the limit $\mathcal{N} \to \infty$, the corresponding probability distribution tends to a Dirac delta function

$$P(x) \equiv \frac{M(x)}{\int_0^{u/2} dy M(y)} \to \delta(x - x_M).$$

This confirms that the probability concentrates on a single distribution (a single x in the present case), which is the most probable one. Even if there are exceptions to this law, for instance in some systems presenting a critical point (like the liquid–vapor critical point), that does not regard the systems considered in this text, so that the equilibrium distribution can be surely identified with the most likely one.

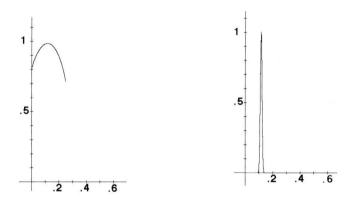

Fig. 3.2. Two plots in arbitrary units of the multiplicity distribution $\mathcal{M}(x)$ of the $\mathcal{N}$-element Gibbs ensemble of the three level model. The comparison of the distribution for $\mathcal{N} = 1$ (left) and $\mathcal{N} = 1000$ (right) shows the generally foreseen fast reduction of the fluctuations for increasing $\mathcal{N}$

In Fig. 3.2 the plots of the function given in (3.39) are shown for an arbitrary choice of the vertical scale. We have set $u = 1/2$ and we show two different cases, $\mathcal{N} = 1$ (left) and $\mathcal{N} = 1000$ (right).

Making always reference to the three-level
system, we notice that the ratios of the occupation numbers are given by

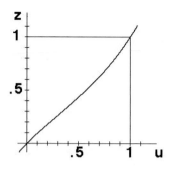

$$z = e^{-\beta\epsilon} = \frac{x_M}{u - 2x_M}$$

$$= \frac{\sqrt{1 + 6u - 3u^2} + u - 1}{4 - 2u} . \quad (3.43)$$

The plot of z as a function of u is given on
the side and shows that in the range $(0, 1)$
we have $0 \le z \le 1$. Hence $\beta \to \infty$ as $u \to 0$
and $\beta \to 0$ as $u \to 1$.

3.4 The Grand Canonical Ensemble
and the Perfect Quantum Gas

We will describe schematically the perfect gas as a system made up of a great
number of atoms, molecules or in general particles of the same species, which
have negligible interactions among themselves but are subject to external
forces. We can consider for instance a gas of particles elastically attracted
towards a fixed point or instead a gas of free particles contained in a box with
reflecting walls. We will show detailed computations for the last case, since it
has several interesting applications, but the reader is invited to think about
the generalization of the results that will be obtained to the case of different
external forces.

The states of every particle in the box can be described as we have done for
the single particle in a box, see Paragraph 3.1.2. However, classifying the states
of many identical particles raises a new problem of quantum nature, which
is linked to *quantum indistinguishability* and to the corresponding statistical
choice.

The uncertainty principle is in contrast with the idea of particle trajectory.
If a particle is located with a good precision at a given time, then its velocity
is highly uncertain and so will be its position at a later time. If two identical
particle are located very accurately at a given time t around points r_1 and r_2,
their positions at later times will be distributed in a quite random way; if we
locate again the particles at time $t + \Delta t$ we could not be able to decide which
of the two particles corresponds to that initially located in r_1 and which to
the other.

The fact that the particles cannot be distinguished implies that the probability density of locating the particles in two given points, $\rho(r_1, r_2)$, must
necessarly be symmetric under exchange of its arguments:

$$\rho(r_1, r_2) = \rho(r_2, r_1) . \quad (3.44)$$

Stated otherwise, indices 1 and 2 refer to the points where the two particles are simultaneously located but in no way identify which particle is located where.

If we consider that also in the case of two (or more) particles the probability density must be linked to the wave functions by the relation:

$$\rho(\boldsymbol{r}_1, \boldsymbol{r}_2) = |\psi(\boldsymbol{r}_1, \boldsymbol{r}_2)|^2 ,$$

then, taking (3.44) into account, we have that:

$$\psi(\boldsymbol{r}_1, \boldsymbol{r}_2) = \mathrm{e}^{\mathrm{i}\phi} \, \psi(\boldsymbol{r}_2, \boldsymbol{r}_1) ,$$

where ϕ cannot depend on positions since that would change the energy of the corresponding state. A double exchange implies that $\mathrm{e}^{2\mathrm{i}\phi} = 1$, so that $\mathrm{e}^{\mathrm{i}\phi} = \pm 1$. Therefore we can state that, in general, the wave function of two identical particles must satisfy the following symmetry relation

$$\psi(\boldsymbol{r}_1, \boldsymbol{r}_2) = \pm \, \psi(\boldsymbol{r}_2, \boldsymbol{r}_1) . \tag{3.45}$$

Since identity among particles is equivalent to the invariance of Schrödinger equation under coordinates exchange, we conclude that equation (3.45) is yet another application of the symmetry principle introduced in Section 2.6.

Generalizing the same argument to the case of more than two particles, it can be easily noticed that the sign appearing in (3.45) must be the same for all identical particles of the same species. The plus sign applies to photons, to hydrogen and helium atoms, to biatomic molecules made up of identical atoms and to many other particles. There is also a large number of particles for which the minus sign must be used, in particular electrons, protons and neutrons. In general, particles of the first kind are called *bosons*, while particles of the second kind are named *fermions*.

As we have seen at the end of Section 2.3, particles may have an internal angular momentum which is called , whose projection ($\hbar s$) in a given direction, e.g. in the momentum direction, can assume the values $(S-m)\hbar$ where m is an integer such that $0 \leq m \leq 2S$ and S is either integer or half-integer. In case of particles with non-vanishing mass one has an independent state for each value of m. This is not true for massless particles. Indeed, e.g. for a photon, the spin momentum projection assumes only two possible values ($\pm\hbar$), corresponding to the independent polarizations of light. A general theorem (*spin-statistics theorem*) states that particles carrying half-integer spin are fermions, while those for which S is an integer are bosons.

Going back to the energy levels of a system made up of two particles in a box with reflecting walls, they are given by

$$E = \frac{\pi^2 \hbar^2}{2mL^2} \left[k_{x,1}^2 + k_{y,1}^2 + k_{z,1}^2 + k_{x,2}^2 + k_{y,2}^2 + k_{z,2}^2 \right] . \tag{3.46}$$

The corresponding states are identified by two vectors (*wave vectors*) $\boldsymbol{k}_1$ and $\boldsymbol{k}_2$ with positive integer components and, in case, by two spin indices s_1 and

s_2. Indeed, as we have already said, the generic state of a particle carrying spin is described by a wave function with complex components which can be indicated with $\psi(\mathbf{r}, \sigma)$, where σ identifies the single component; in this case $|\psi(\mathbf{r}, \sigma)|^2$ represents the probability density of finding the particle around $\mathbf{r}$ and in the spin state σ.

Indicating by $\psi_{\mathbf{k}}(\mathbf{r})$ the wave function of a single particle in a box given in (2.99), the total wave function for two particles, which we assume to have well definite spin components s_1 and s_2, should correspond to the product $\psi_{\mathbf{k}_1}(\mathbf{r}_1)\psi_{\mathbf{k}_2}(\mathbf{r}_2)\delta_{s_1,\sigma_1}\delta_{s_2,\sigma_2}$, but (3.45) compels us to (anti-)symmetrize the wave function, which can then be written as:

$$\psi(\mathbf{r}_1, \mathbf{r}_2, \sigma_1, \sigma_2) = N\big[\ \psi_{\mathbf{k}_1}(\mathbf{r}_1)\psi_{\mathbf{k}_2}(\mathbf{r}_2)\delta_{\sigma_1,s_1}\delta_{\sigma_2,s_2}$$
$$\pm\ \psi_{\mathbf{k}_2}(\mathbf{r}_1)\psi_{\mathbf{k}_1}(\mathbf{r}_2)\delta_{\sigma_1,s_2}\delta_{\sigma_2,s_1}\big]. \qquad (3.47)$$

However that leads to a paradox in case the two wave vectors coincide, $\mathbf{k}_1 = \mathbf{k}_2$, and the particles are fermions in the same spin state, $s_1 = s_2$. Indeed in this case the minus sign has to be used in (3.47), leading to a vanishing result. The only possible solution to this seeming paradox is *Pauli's Exclusion Principle*, which forbids the presence of two identical fermions having the same quantum numbers (wave vector and spin in the present example).

The identification of the states of two particles which can be obtained by exchanging both wave vectors and spin states suggests that a better way to describe them, alternative to fixing the quantum numbers of the single particles, is that of indicating which combinations (wave vector, spin state) appear in the total wave function, and in case of bosons how many times do they appear: that corresponds to indicating which single particle states (each identified by $\mathbf{k}$ and σ) are occupied and by how many particles. In conclusion, the microscopic state of system made up of many identical particles (quantum gas) can be described in terms of the *occupation numbers* of the quantum states accessible to a single particle: they can be non-negative integers in the case of bosons, while only two possibilities, 0 or 1, are left for fermions. For instance the wave function in (3.47) can be described in terms of the following occupation numbers:

$$n_{\mathbf{k},\sigma} = \delta_{\mathbf{k},\mathbf{k}_1}\delta_{\sigma,s_1} + \delta_{\mathbf{k},\mathbf{k}_2}\delta_{\sigma,s_2}.$$

3.4.1 The Perfect Fermionic Gas

According to what stated above, we will consider a system made up of n identical and non-interacting particles of spin $S = 1/2$ constrained in a cubic box with reflecting walls and an edge of length L. Following Gibbs, the box is supposed to be in thermal contact with $\mathcal{N}$ identical boxes.

The generic microscopic state of the gas, which is indicated with an index i in Gibbs construction, is assigned once the occupation numbers $\{n_{\mathbf{k},s}\}$ are given (with $\{n_{\mathbf{k},s}\} = 0$ or 1) for every value of the wave vector $\mathbf{k}$ and of the spin projection $s = \pm 1/2$, with the obvious constraint:

$$\sum_{\mathbf{k},s} n_{\mathbf{k},s} = n .$$ (3.48)

The corresponding energy is given by:

$$E_{\{n_{\mathbf{k},s}\}} = \sum_{\mathbf{k},s} \frac{\hbar^2 \pi^2}{2mL^2} \left(k_x^2 + k_y^2 + k_z^2 \right) n_{\mathbf{k},s} \equiv \sum_{\mathbf{k},s} \frac{\hbar^2 \pi^2}{2mL^2} k^2 n_{\mathbf{k},s} .$$ (3.49)

Notice that all occupation numbers $n_{\mathbf{k},s}$ refer to the particles present in the specified single particle state: they must not be confused with the numbers describing the distribution of the $\mathcal{N}$ copies of the system in Gibbs method. The partition function of our gas is therefore:

$$
\begin{aligned}
Z &= \sum_{\{n_{\mathbf{k},s}\}:\sum_{\mathbf{k},s} n_{\mathbf{k},s}=n} e^{-\beta E_{\{n_{\mathbf{k},s}\}}} = \sum_{\{n_{\mathbf{k},s}\}:\sum_{\mathbf{k},s} n_{\mathbf{k},s}=n} e^{-\beta \sum_{\mathbf{k},s} \frac{\hbar^2 \pi^2}{2mL^2} k^2 n_{\mathbf{k},s}} \\
&= \sum_{\{n_{\mathbf{k},s}\}:\sum_{\mathbf{k},s} n_{\mathbf{k},s}=n} \prod_{\mathbf{k},s} e^{-\beta \frac{\hbar^2 \pi^2}{2mL^2} k^2 n_{\mathbf{k},s}} .
\end{aligned}
$$ (3.50)

The constraint in (3.48) makes the computation of the partition function really difficult. Indeed, without that constraint, last sum in (3.50) would factorize in the product of the different sums over the occupation numbers of the single particle states $\mathbf{k}, s$.

This difficulty can be overcome by relaxing the constraint on the number of particles in each system, keeping only that on the total number n_{tot} of particles in the macrosystem, similarly to what has been done for the energy. Hence the number of particles in the gas, n, will be replaced by the average number $\bar{n} \equiv n_{tot}/\mathcal{N}$. Also in this case the artifice works well, since the probability of the various possible distributions of the macrosystem is extremely peaked around the most likely distribution, so that the number of particles in each systems has negligible fluctuations with respect to the average number $\bar{n}$.

This artifice is equivalent to replacing the Canonical Distribution by the *Grand Canonical* one. In practice, the reflecting walls of our systems are given a small permeability, so that they can exchange not only energy but also particles. In the general case the Grand Canonical Distribution refers to systems made up of several different particle species, however we will consider the case of a single species in our computations.

Going along the same lines of the construction given in Section 3.1, we notice that the generic state of the system under consideration, identified by the index i, is now characterized by its particle number n_i as well as by its energy E_i. We are therefore looking for the distribution having maximum multiplicity

$$M \left(\{N_i\} \right) = \frac{\mathcal{N}!}{\prod_i N_i!} ,$$

constrained by the total number of considered systems

$$\sum_i N_i = \mathcal{N} , \tag{3.51}$$

by the total energy of the macrosystem

$$\sum_i N_i E_i = U\mathcal{N} \tag{3.52}$$

and, as an additional feature of the Grand Canonical Distribution, by the total number of particles of each species. In our case, since we are dealing with a single type of particles, there is only one additional constraint:

$$\sum_i n_i N_i = \bar{n} \, \mathcal{N} . \tag{3.53}$$

If we apply the method of Lagrange's multipliers we obtain, analogously to what has been done for the Canonical Distribution, see (3.12),

$$\ln \, p_i = -1 + \gamma - \beta \left(E_i - \mu n_i \right) , \tag{3.54}$$

where we have introduced the new Lagrange multiplier $\beta\mu$, associated with the constraint in (3.53). We arrive finally, in analogy with the canonical case, to the probability in the Grand Canonical Distribution:

$$p_i = \frac{e^{-\beta(E_i - \mu n_i)}}{\sum_j e^{-\beta(E_j - \mu n_j)}} \equiv \frac{e^{-\beta(E_i - \mu n_i)}}{\Xi} , \tag{3.55}$$

where the grand canonical partition function, Ξ, has been implicitly defined.

It can be easily shown that, in the same way as energy exchange (thermal equilibrium) compels the Lagrange multiplier β to be the same for all systems in thermal contact (that has been explicitly shown for the Canonical Distribution), particle exchange forces all systems to have the same *chemical potential* μ for each particle species separately. The chemical potential can be computed through the expression for the average number of particles:

$$\bar{n} = \sum_i n_i p_i = \sum_i n_i \frac{e^{-\beta(E_i - \mu n_i)}}{\Xi} . \tag{3.56}$$

Let us now go back to the case of the perfect fermionic gas. The Grand Canonical partition function can be written as:

$$\Xi = \sum_{\{n_{\boldsymbol{k},s}\}} e^{-\beta\left(E_{\{n_{\boldsymbol{k},s}\}} - \mu \sum_{\boldsymbol{k},s} n_{\boldsymbol{k},s} \right)} = \sum_{\{n_{\boldsymbol{k},s}\}} e^{-\beta \sum_{\boldsymbol{k},s} \left(\frac{\hbar^2 \pi^2}{2mL^2} k^2 - \mu \right) n_{\boldsymbol{k},s}}$$

$$= \prod_{\boldsymbol{k},s} \left(\sum_{n_{\boldsymbol{k},s}=0}^{1} e^{-\beta\left(\frac{\hbar^2 \pi^2}{2mL^2} k^2 - \mu \right) n_{\boldsymbol{k},s}} \right) = \prod_{\boldsymbol{k},s} \left(1 + e^{-\beta\left(\frac{\hbar^2 \pi^2}{2mL^2} k^2 - \mu \right)} \right) \tag{3.57}$$

Hence, based on (3.55), we can write the probability of the state defined by the occupation numbers $\{n_{k,s}\}$ as

$$p\left(\{n_{k,s}\}\right) = \frac{e^{-\beta \sum_{k,s}\left(\frac{\hbar^2\pi^2k^2}{2mL^2}-\mu\right)n_{k,s}}}{\Xi} = \prod_{k,s}\left(\frac{e^{-\beta n_{k,s}\left(\frac{\hbar^2\pi^2k^2}{2mL^2}-\mu\right)}}{1+e^{-\beta\left(\frac{\hbar^2\pi^2k^2}{2mL^2}-\mu\right)}}\right), \quad (3.58)$$

which, as can be easily seen, factorizes into the product of the probabilities related to the single particle states: $p\left(\{n_{k,s}\}\right) = \prod_{k,s} p\left(n_{k,s}\right)$, where

$$p\left(n_{k,s}\right) = \frac{e^{-\beta n_{k,s}\left(\frac{\hbar^2\pi^2k^2}{2mL^2}-\mu\right)}}{1+e^{-\beta\left(\frac{\hbar^2\pi^2k^2}{2mL^2}-\mu\right)}}. \quad (3.59)$$

Using this result we can easily derive the average occupation number for the single particle state, also known as *Fermi–Dirac distribution*:

$$\bar{n}_{k,s} = \sum_{n_{k,s}=0}^{1} n_{k,s}p\left(n_{k,s}\right) = \frac{e^{-\beta\left(\frac{\hbar^2\pi^2k^2}{2mL^2}-\mu\right)}}{1+e^{-\beta\left(\frac{\hbar^2\pi^2k^2}{2mL^2}-\mu\right)}} = \frac{1}{1+e^{\beta\left(\frac{\hbar^2\pi^2k^2}{2mL^2}-\mu\right)}}. \quad (3.60)$$

This result can be easily generalized to the case of fermions which are subject to an external force field, leading to single particle energy levels E_α identified by one or more indices α. The average occupation number of the single particle state α is then given by:

$$\bar{n}_\alpha = \frac{1}{1+e^{\beta(E_\alpha-\mu)}}. \quad (3.61)$$

and the chemical potential can be computed making use of (3.56), which according to (3.61) and (3.60) can be written in the following form:

$$\bar{n} = \sum_\alpha \frac{1}{1+e^{\beta(E_\alpha-\mu)}} = \sum_{k,s} \frac{1}{1+e^{\beta\left(\frac{\hbar^2\pi^2k^2}{2mL^2}-\mu\right)}}, \quad (3.62)$$

where the second equation is valid for free particles.

We have therefore achieved a great simplification in the description of our system by adopting the Grand Canonical construction. This simplification can be easily understood in the following terms. Having relaxed the constraint on the total number of particles in each system, each single particle state can be effectively considered as an independent sub-system making up, together with all other single particle states, the whole system. Each single particle sub-system can be found, for the case of fermions, in only two possible states with occupation number 0 or 1: its Grand Canonical partition function is therefore trivially given by $1 + \exp(-\beta(E_\alpha - \mu))$, with β and μ being the same for all sub-systems because of thermal and chemical equilibrium. The probability distribution, equation (3.59), and the Fermi–Dirac distribution easily follows.

In order to make use of previous formulae, it is convenient to arrange the single particle states k, s according to their energy, thus replacing the sum over state indices by a sum over state energies. With that aim, let us recall that the possible values of k, hence the possible states, correspond to the vertices of a cubic lattice having spacing of length 1. In the nearby figure we show the lattice for the two-dimensional case. It is clear that, apart from small corrections due to the discontinuity in the distribution of vertices, the number of single particle states having energy less than a given value E is given by

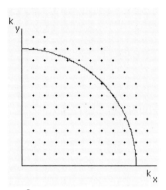

$$n_E = \frac{2}{8} \frac{4\pi \left(\frac{\sqrt{2mEL}}{\pi\hbar} \right)^3}{3} = \frac{\left(\frac{\sqrt{2mL}}{\hbar} \right)^3}{3\pi^2} E^{3/2} , \tag{3.63}$$

which is equal to the volume of the sphere of radius

$$k = \frac{\sqrt{2mEL}}{\pi\hbar}$$

divided by the number of sectors (which is 8 in three dimensions), since k has only positive components, and multiplied by the number of spin states, e.g. 2 for electrons.

The approximation used above, which improves at fixed particle density $\bar{n}/L^3$ as the volume L^3 increases, consists in considering the single particle states as distributed as a function of their energy in a continuous, instead of discrete, way. On this basis, we can compute the density of single particle states as a function of energy:

$$\frac{dn_E}{dE} = \frac{\sqrt{2m^3}L^3}{\pi^2\hbar^3} \sqrt{E} . \tag{3.64}$$

Hence we can deduce from (3.60) the distribution of particles as a function of their energy:

$$\frac{d\bar{n}(E)}{dE} = \frac{\sqrt{2m^3}L^3}{\pi^2\hbar^3} \frac{\sqrt{E}}{1 + e^{\beta(E-\mu)}} . \tag{3.65}$$

and replace (3.62) by the following equation:

$$\bar{n} = \int_0^\infty \frac{d\bar{n}(E)}{dE} dE = \int_0^\infty dE \frac{\sqrt{2m^3}L^3}{\pi^2\hbar^3} \frac{\sqrt{E}}{1 + e^{\beta(E-\mu)}} . \tag{3.66}$$

Equation (3.65) has a simple interpretation in the limit $T \to 0$, i.e. as $\beta \to \infty$. Indeed, in that limit, the exponential in the denominator diverges for all single particle states having energy greater than μ, hence the occupation number vanishes for those states. The exponential instead vanishes for states having energy less that μ, for which the occupation number is one. Therefore the chemical potential in the limit of low temperatures, which is also called *Fermi energy E_F*, can be computed by the equation:

$$n_{E_F} = \frac{\left(\frac{\sqrt{2m}L}{\hbar}\right)^3}{3\pi^2} E_F^{3/2} = \bar{n} \cdot . \tag{3.67}$$

Solving for E_F, we obtain:

$$E_F = \mu|_{T=0} = \frac{\hbar^2}{2m}\left(3\pi^2\rho\right)^{2/3} , \tag{3.68}$$

where $\rho = \bar{n}/L^3$ is the density of particles in the gas.

In order to discuss the opposite limit, in which T is very large ($\beta \to 0$), let us set $z = e^{-\beta\mu}$ and rewrite (3.66) by changing the integration variable ($x = \beta E$):

$$\bar{n} = \frac{\sqrt{2m^3}L^3}{\pi^2\hbar^3\beta^{\frac{3}{2}}} \int_0^\infty dx \frac{\sqrt{x}}{1 + ze^x} = \frac{\sqrt{2(mkT)^3}L^3}{\pi^2\hbar^3} \int_0^\infty dx \frac{\sqrt{x}}{1 + ze^x} . \tag{3.69}$$

We see from this equation that μ must tend to $-\infty$ as $T \to \infty$, i.e. z must diverge, otherwise the right-hand side in (3.69) would diverge like $T^{3/2}$, which is a nonsense since $\bar{n}$ is fixed a priori.

Since at high temperatures z diverges, the exponential in the denominator of (3.60) is much greater than 1, hence (3.65) can be replaced, within a good approximation, by

$$\frac{d\bar{n}(E)}{dE} = \frac{\left(\frac{\sqrt{2m}L}{\hbar}\right)^3}{2\pi^2 z} \sqrt{E}e^{-\beta E} \equiv AL^3\sqrt{E}e^{-\beta E} . \tag{3.70}$$

The constant A, hence μ, can be computed through

$$\int_0^\infty AL^3\sqrt{E}e^{-\beta E}dE = 2AL^3 \int_0^\infty x^2 e^{-\beta x^2}dx = -2AL^3 \frac{d}{d\beta} \int_0^\infty e^{-\beta x^2}dx$$

$$= -AL^3 \frac{d}{d\beta} \int_{-\infty}^\infty e^{-\beta x^2}dx = -AL^3 \frac{d}{d\beta}\sqrt{\frac{\pi}{\beta}} = \frac{AL^3}{2}\sqrt{\frac{\pi}{\beta^3}} = \bar{n} . \tag{3.71}$$

We have therefore $A = 2\rho\sqrt{\beta^3/\pi} = e^{\beta\mu}\sqrt{2m^3}/(\pi^2\hbar^3)$, confirming that $\mu \to -\infty$ as $\beta \to 0$ ($\mu \sim \ln\beta/\beta$).

It is remarkable that in the limit under consideration, in which the distribution of particles according to their energy is given by:

$$\frac{d\bar{n}(E)}{dE} = 2\rho\sqrt{\frac{\beta^3}{\pi}}L^3\sqrt{E}e^{-\beta E} , \tag{3.72}$$

Planck's constant has disappeared from formulae. If consequently we adopt the classical formula for the energy of the particles, $E = mv^2/2$, we can find the velocity distribution corresponding to (3.70):

$$\frac{d\bar{n}(v)}{dv} \equiv \frac{d\bar{n}(E)}{dE}\frac{dE}{dv} = \rho\sqrt{\frac{2m^3\beta^3}{\pi}}L^3 v^2 e^{-\beta mv^2/2}. \tag{3.73}$$

Replacing β by $1/kT$, equation (3.73) reproduces the well known Maxwell distribution for velocities, thus confirming the identification $\beta = 1/kT$ made before.

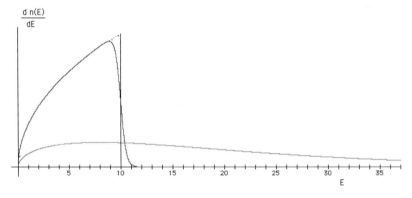

Fig. 3.3. A plot of the Fermi–Dirac energy distribution in arbitrary units for a fermion gas with $E_F = 10$ for $kT = 0$, 0.25 and 12.5. Notice the for the first two values of kT the distribution saturates the Pauli exclusion principle limit, while for $kT = 12.5$ it approaches the Maxwell–Boltzmann distribution

The figure above reproduces the behavior predicted by (3.65) for three different values of kT, and precisely for $kT = 0$, 0.25 and 12.5, measured in the arbitrary energy scale given in the figure, according to which $E_F = 10$. The two curves corresponding to lower temperatures show saturation for states with energy $E < E_F$, in contrast with the third curve which instead reproduces part of the Maxwell distribution and corresponds to small occupation numbers.

Making use of (3.65) we can compute the mean energy U of the gas:

$$U = \int_0^\infty E\frac{d\bar{n}(E)}{dE}dE = \frac{\sqrt{2m^3}L^3}{\pi^2\hbar^3}\int_0^\infty \frac{\sqrt{E^3}}{1 + e^{\beta(E-\mu)}}dE, \tag{3.74}$$

obtaining, in the low temperature limit,

$$U = \frac{\sqrt{8m^3}L^3}{5\pi^2\hbar^3}E_F^{5/2} = \frac{(3\bar{n})^{5/3}\pi^{4/3}\hbar^2}{10mL^2} = \frac{(3\bar{n})^{5/3}\pi^{4/3}\hbar^2}{10mV^{2/3}}, \tag{3.75}$$

where $V = L^3$ is the volume occupied by the gas. At high temperatures we have instead:

$$U = 2\rho\sqrt{\frac{\beta^3}{\pi}}L^3 \int_0^\infty E^{3/2}e^{-\beta E}dE = 4\rho\sqrt{\frac{\beta^3}{\pi}}L^3 \int_0^\infty x^4 e^{-\beta x^2}dx$$

$$= \frac{3}{2}\bar{n}\sqrt{\frac{\beta^3}{\pi}}\sqrt{\frac{\pi}{\beta^5}} = \frac{3}{2\beta}\bar{n} \equiv \frac{3}{2}\bar{n}kT. \tag{3.76}$$

This result reproduces what predicted by the classical kinetic theory and in particular the specific heat at constant volume for a gram atom of gas: $C_V = 3/2kN_A \equiv 3/2R$.

In order to compute the specific heat in the low temperature case, we notice first that, for large values of β, equation (3.66) leads, after some computations, to $\mu = E_F - O(\beta^{-2})$, hence $\mu = E_F$ also at the first order in T. We can derive (3.74) with respect to T, obtaining:

$$C = k\beta^2 \frac{\sqrt{2m^3}L^3}{\pi^2\hbar^3} \int_0^\infty \frac{\sqrt{E^3}(E-\mu)\,e^{\beta(E-\mu)}}{(1+e^{\beta(E-\mu)})^2}dE. \tag{3.77}$$

For large values of β the exponential factor in the numerator makes the contributions to the integral corresponding to $E \ll \mu$ negligible, while the exponential in the denominator makes negligible contributions from $E \gg \mu$. This permits to make a Taylor expansion of the argument of the integral in (3.77). In particular, if we want to evaluate contributions of order T, taking (3.67) into account we obtain:

$$C \simeq k\beta^2 \frac{3\bar{n}}{8E_F^{\frac{3}{2}}} \int_{-\infty}^\infty \frac{\sqrt{E^3}(E-E_F)}{\left(\cosh\frac{\beta(E-E_F)}{2}\right)^2}dE \simeq k\beta^2 \frac{3\bar{n}}{8} \int_{-\infty}^\infty \frac{(E-E_F)}{\left(\cosh\frac{\beta(E-E_F)}{2}\right)^2}dE$$

$$+ k\beta^2 \frac{9\bar{n}}{16E_F} \int_{-\infty}^\infty \frac{(E-E_F)^2}{\left(\cosh\frac{\beta(E-E_F)}{2}\right)^2}dE = k\bar{n}\frac{9kT}{2E_F} \int_{-\infty}^\infty \left(\frac{x}{\cosh x}\right)^2 dx$$

$$= k\bar{n}\frac{3\pi^2 kT}{4E_F}, \tag{3.78}$$

showing that the specific heat has a linear dependence in T at low temperatures.

The linear growth of the specific heat with T for low temperatures can be easily understood in terms of the distribution of particles at low T. In particular we can make reference to (3.65) and to its graphical representation shown in Fig. 3.3: at low temperatures the particles occupy the lowest possible energy levels, thus saturating the limit imposed by Pauli's principle. In these conditions most of the particles cannot exchange energy with the external environment, since energy exchanges of the order of kT, which are typical at temperature T, would imply transitions of a particle to a different energy level which however is already completely occupied by other particles. If we refer to the curve corresponding to $kT = 0.25$ in the figure, we see that only particles having energies in a small interval of width kT around the Fermi energy ($E_F = 10$ in the figure), where the occupation number rapidly goes from 1 to 0, can make transitions from one energy level to another, thus exchanging energy with the reservoir and giving a contribution to the specific heat, which is of the order of k for each particle. We expect therefore a specific heat which is reduced by a factor kT/E_F with respect to that for the high temperature case: this roughly reproduces the exact result given in (3.78).

Evidently we expect the results we have obtained to apply in particular to the electrons in the conduction band of a metal. It could seem that a gas of electrons be far from being non-interacting, since electrons exchange repulsive Coulomb interactions; however Coulomb forces are largely screened by the other charges present in the metallic lattice, and can therefore be neglected, at least qualitatively, at low energies. We recall that in metals there is one free electron for each atom, therefore, taking iron as an example, which has a mass density $\rho_m \simeq 5 \, 10^3$ Kg/m^3 and atomic weight $A \simeq 50$, the electronic density is: $\bar{n}/V = \rho_m(N_A/A)10^3 \sim 6 \, 10^{28}$ particles/m^3. Making use of (3.68) we obtain: $E_F \simeq (10^{-68}/1.8 \, 10^{-30}) \, (3\pi^2 6 \, 10^{28})^{2/3}$ J $\simeq 10^{-18}$ J $\simeq 6$ eV. If we recall that at room temperature $kT \simeq 2.5 \, 10^{-2}$ eV, we see that the order of magnitude of the contribution of electrons to the specific heat is $3 \, 10^{-2}R$ per gram atom, to be compared with $3R$, which is the contribution of atoms according to Dulong and Petit. Had we not taken quantum effects into account, thus applying the equipartition principle, we would have predicted a contribution $3/2R$ coming from electrons. That gives a further confirmation of the relevance of quantum effects for electrons in matter.

Going back to the general case, we can obtain the equation of state for a fermionic gas by using (3.35). From the expression for the energy E_i corresponding to a particular state of the gas given in (3.49) we derive $\partial E_i/\partial V = -2E_i/3V$, hence

$$PV = \frac{2}{3}U, \qquad (3.79)$$

which at high temperatures, where (3.76) is valid, reproduces the well known perfect gas law. Notice that the equation of state in the form given in (3.79) is identical to that obtained in (3.33) and indeed depends only on the dispersion relation (giving energy in terms of momentum) assumed for the free particle states, i.e. the simple form and the factor 2/3 are strictly related to having considered the particles as non-relativistic (see Problem 3.22 for the case of ultrarelativistic particles).

3.4.2 The Perfect Bosonic Gas

To complete our program, let us consider a gas of spinless atoms, hence of bosons; in order to have a phenomenological reference, we will think in particular of a mono-atomic noble gas like helium. The system can be studied along the same lines followed for the fermionic gas, describing its possible states by assigning the occupation numbers of the single particle states, with the only difference that in this case the wave function must be symmetric in its arguments (the coordinates of the various identical particles), hence there is no limitation on the number of particles occupying a given single particle state. The Grand Canonical partition function for a bosonic gas in a box is therefore

$$
\Xi = \sum_{\{n_{\boldsymbol{k}}\}} e^{-\beta\left(E_{\{n_{\boldsymbol{k}}\}} - \mu \sum_{\boldsymbol{k}} n_{\boldsymbol{k}}\right)} = \sum_{\{n_{\boldsymbol{k}}\}} e^{-\beta \sum_{\boldsymbol{k}} \left(\frac{\hbar^2 \pi^2}{2mL^2} k^2 - \mu\right) n_{\boldsymbol{k}}}
$$

$$
= \prod_{\boldsymbol{k}} \left(\sum_{n_{\boldsymbol{k}}=0}^{\infty} e^{-\beta\left(\frac{\hbar^2 \pi^2}{2mL^2} k^2 - \mu\right) n_{\boldsymbol{k}}} \right)
$$

$$
= \prod_{\boldsymbol{k}} \frac{1}{1 - e^{-\beta\left(\frac{\hbar^2 \pi^2}{2mL^2} k^2 - \mu\right)}} \, , \tag{3.80}
$$

from which the probability of the generic state of the gas follows:

$$
p(\{n_{\boldsymbol{k}}\}) = \frac{e^{-\beta \sum_{\boldsymbol{k}} \left(\frac{\hbar^2 \pi^2 k^2}{2mL^2} - \mu\right) n_{\boldsymbol{k}}}}{\Xi}
$$

$$
= \prod_{\boldsymbol{k}} \left(e^{-\beta n_{\boldsymbol{k}} \left(\frac{\hbar^2 \pi^2 k^2}{2mL^2} - \mu\right)} \left(1 - e^{-\beta\left(\frac{\hbar^2 \pi^2 k^2}{2mL^2} - \mu\right)}\right) \right), \tag{3.81}
$$

which again can be written as the product of the occupation probabilities relative to each single particle state:

$$
p(n_{\boldsymbol{k}}) = e^{-\beta n_{\boldsymbol{k}} \left(\frac{\hbar^2 \pi^2 k^2}{2mL^2} - \mu\right)} \left(1 - e^{-\beta\left(\frac{\hbar^2 \pi^2 k^2}{2mL^2} - \mu\right)}\right). \tag{3.82}
$$

The average occupation number of the generic single particle state, $\boldsymbol{k}$, is thus

$$
\bar{n}_{\boldsymbol{k}} = \sum_{n_{\boldsymbol{k}}=0}^{\infty} n_{\boldsymbol{k}} \, p(n_{\boldsymbol{k}}) = \left(1 - e^{-\beta\left(\frac{\hbar^2 \pi^2 k^2}{2mL^2} - \mu\right)}\right) \sum_{n=0}^{\infty} n \, e^{-\beta n \left(\frac{\hbar^2 \pi^2 k^2}{2mL^2} - \mu\right)}
$$

$$
= \frac{e^{-\beta\left(\frac{\hbar^2 \pi^2 k^2}{2mL^2} - \mu\right)}}{1 - e^{-\beta\left(\frac{\hbar^2 \pi^2 k^2}{2mL^2} - \mu\right)}} = \frac{1}{e^{\beta\left(\frac{\hbar^2 \pi^2 k^2}{2mL^2} - \mu\right)} - 1} \, , \tag{3.83}
$$

which is known as the *Bose–Einstein* distribution. We deduce from last equation that the chemical potential cannot be greater than the energy of the fundamental single particle state, i.e.

$$
\mu \le 3 \frac{\hbar^2 \pi^2}{2mL^2} \, ,
$$

otherwise the average occupation number of that state would be negative. In the limit of large volumes the fundamental state for a particle in a box has vanishing energy, hence μ must be negative.

The exact value of the chemical potential is fixed by the relation

$$\bar{n} = \sum_{k} \bar{n}_k = \sum_{k} \frac{1}{e^{\beta \left(\frac{\hbar^2 \pi^2 k^2}{2mL^2} - \mu \right)} - 1} \, . \tag{3.84}$$

For the explicit computation of μ we can make use, as in the fermionic case, of the distribution in energy, considering it approximately as a continuous function:

$$\frac{d\bar{n}(E)}{dE} = \sqrt{\frac{m^3}{2}} \frac{L^3}{\pi^2 \hbar^3} \frac{\sqrt{E}}{e^{\beta(E-\mu)} - 1} \, . \tag{3.85}$$

Notice that equation (3.85) differs from the analogous given in (3.65), which is valid in the fermionic case, both for the sign in the denominator and for a global factor $1/2$ which is due to the absence of the spin degree of freedom.

The continuum approximation for the distribution of the single particle states in energy is quite rough for small energies, where only few levels are present. In the fermionic case, however, that is not a problem, since, due to the Pauli exclusion principle, only a few particles can occupy those levels (2 per level at most in the case of electrons), so that the contribution coming from the low energy region is negligible. The situation is quite different in the bosonic case. If the chemical potential is small, the occupation number of the lowest energy levels can be very large, giving a great contribution to the sum in (3.84). We exclude for the time being this possibility and compute the chemical potential making use of the relation:

$$\bar{n} = \sqrt{\frac{m^3}{2}} \frac{L^3}{\pi^2 \hbar^3} \int_0^\infty dE \frac{\sqrt{E}}{e^{\beta(E-\mu)} - 1} = \sqrt{\frac{(mkT)^3}{2}} \frac{L^3}{\pi^2 \hbar^3} \int_0^\infty dx \frac{\sqrt{x}}{z e^x - 1} \tag{3.86}$$

where again we have set $z = e^{-\beta\mu} \geq 1$. Defining the gas density, $\rho \equiv \bar{n}/L^3$, equation (3.86) can be rewritten as:

$$\int_0^\infty dx \frac{\sqrt{x}}{z e^x - 1} = \pi^2 \hbar^3 \rho \sqrt{\frac{2}{(mkT)^3}} \, . \tag{3.87}$$

On the other hand, recalling that $z \geq 1$, we obtain the following inequality:

$$\int_0^\infty dx \frac{\sqrt{x}}{z e^x - 1} \leq \frac{1}{z} \int_0^\infty dx \frac{\sqrt{x}}{e^x - 1} \leq \int_0^\infty dx \frac{\sqrt{x}}{e^x - 1} \simeq 2.315 \, , \tag{3.88}$$

which can be replaced in (3.87), giving an upper limit on the ratio $\rho/T^{3/2}$. That limit can be interpreted as follows: for temperatures lower than a certain threshold, the continuum approximation for the energy levels cannot account for the distribution of all particles in the box, so that we must admit a macroscopic contribution coming from the lowest energy states, in particular from the fundamental state. The limiting temperature can be considered as a critical temperature, and the continuum approximation is valid only if

$$T \geq T_c \simeq 4.38 \frac{\hbar^2 \rho^{2/3}}{m \; k} \, . \tag{3.89}$$

As T approaches the critical temperature the chemical potential vanishes and the occupation of the fundamental state becomes comparable with $\bar{n}$, hence of macroscopic nature. For temperatures lower than T_c the computation of the total number of particles shown in (3.86) must be rewritten as:

$$\bar{n} = \bar{n}_f + \sqrt{\frac{m^3}{2}} \frac{L^3}{\pi^2 \hbar^3} \int_0^\infty dE \frac{\sqrt{E}}{e^{\beta E} - 1} \, , \tag{3.90}$$

where $\bar{n}_f$ refers to the particles occupying the lowest energy states, while the integral over the continuum distribution, in which μ has been neglected, takes into account particles occupying higher energy levels. Changing variables in the integral we obtain:

$$\bar{n} \simeq \bar{n}_f + 2.315 \sqrt{\frac{(mkT)^3}{2}} \frac{L^3}{\pi^2 \hbar^3} \, , \tag{3.91}$$

showing that, for $T < T_c$, $\bar{n}_f$ takes macroscopic values, of the order of magnitude of Avogadro's number N_A.

This phenomenon is known as Bose–Einstein condensation. Actually, for the usual densities found in ordinary gasses in normal conditions, i.e. $\rho \simeq 10^{25}$ particles/m^3 , the critical temperature is of the order of 10^{-2} °K, a value at which interatomic forces are no more negligible even in the case of helium, so that the perfect gas approximation does not apply. The situation can be completely different at very low densities, indeed Bose–Einstein condensation has been recently observed for temperatures of the order of 10^{-9} °K and densities of the order of 10^{15} particles/m^3 ,

In the opposite situation, for temperatures much greater than T_c, the exponential clearly dominates in the denominator of the continuum distribution since, analogously to what happens for fermions at high temperatures, one can show that $z \gg 1$. Hence the -1 term can be neglected, so that the distribution becomes that obtained also in the fermionic case at high temperatures, i.e. the Maxwell distribution.

3.4.3 The Photonic Gas and the Black Body Radiation

We will consider in brief the case of an electromagnetic field in a box with "almost" completely reflecting walls: we have to give up ideal reflection in order to allow for thermal exchanges with the reservoir. From the classical point of view, the field amplitude can be decomposed in normal oscillation modes corresponding to well defined values of the frequency and to electric and magnetic fields satisfying the well known reflection conditions on the box surface. The modes under consideration, apart from the two possible polarizations of the electric field, are completely analogous to the wave functions of a particle in a box shown in (2.99), i.e. sinusoidal functions whose argument, choosing the origin of coordinates in a vertex of the box, is equal to

$$\frac{\pi}{L}\left(k_x x + k_y y + k_z z - \sqrt{k_x^2 + k_y^2 + k_z^2}\, ct\right) \equiv \frac{\pi}{L}\left(\mathbf{k}\cdot\mathbf{r} - kct\right)\,, \qquad (3.92)$$

where, as usual, the vector $\mathbf{k}$ has integer components (k_x, k_y, k_z). We have therefore the following frequencies:

$$\nu_{\mathbf{k}} = \frac{c}{2L}\sqrt{k_x^2 + k_y^2 + k_z^2}\,. \qquad (3.93)$$

Taking into account the two possible polarizations, the number of modes having frequency less than a given value ν is:

$$n_\nu = \frac{\pi}{3}|\mathbf{k}|^3 = \frac{8\pi L^3 \nu^3}{3c^3}\,, \qquad (3.94)$$

from which the density of modes can be deduced:

$$\frac{dn_\nu}{d\nu} = \frac{8\pi L^3 \nu^2}{c^3}\,. \qquad (3.95)$$

If the system is placed at thermal equilibrium at a temperature T and we assume equipartition of energy, i.e. that an average energy kT corresponds to each oscillation mode, we arrive to the result found by Rayleigh and Jeans for the energy distribution of the black body[2] radiation as a function of frequency (a quantity which can also be easily measured in the case of an oven):

$$\frac{dU(\nu)}{d\nu} = \frac{8\pi kT}{c^3}L^3\nu^2\,. \qquad (3.96)$$

This is clearly a parodoxical result, since, integrating over frequencies, we would obtain an infinite internal energy, hence an infinite specific heat. From a historical point view it was exactly this paradox which urged Planck to formulate his hypothesis about quantization of energy, which was then better specified by Einstein who assumed the existence of photons.

Starting from Einstein's hypothesis, equation (3.95) can be interpreted as the density of states for a gas of photons, i.e. bosons with energy $E = h\nu$. The density of photons given in (3.64) becomes then:

$$\frac{dn_E}{dE} = \left(\frac{L}{\hbar c}\right)^3 \frac{E^2}{\pi^2}\,. \qquad (3.97)$$

---

[2] A black body, extending a notion valid for the visible electromagnetic radiation, is defined as an ideal body which is able to emit and absorb elettromagnetic radiation of any frequency, so that all oscillation modes interacting with (emitted by) a black body at thermal equilibrium at temperature T can be considered as thermalized at the same temperature.

For a gas of photons the collisions with the walls of the box, which thermalize the system, correspond in practice to non-ideal reflection processes in which photons can absorbed or new photons can be emitted by the walls. Therefore, making always reference to a macrosystem made up of a large number of similar boxes, there is actually no constraint on the total number of particles, hence the chemical potential must vanish.

The distribution law of the photons in energy is then given by:

$$\frac{dn(E)}{dE} = \left(\frac{L}{\hbar c}\right)^3 \frac{E^2}{\pi^2} \frac{1}{e^{\frac{E}{kT}} - 1}. \tag{3.98}$$

From this law we can deduce the distribution in frequency:

$$\frac{dn(\nu)}{d\nu} = \left(\frac{L}{\hbar c}\right)^3 \frac{(h\nu)^2}{\pi^2} \frac{h}{e^{\frac{h\nu}{kT}} - 1} = \frac{8\pi}{c^3} L^3 \frac{\nu^2}{e^{\frac{h\nu}{kT}} - 1} \tag{3.99}$$

and we can finally write the energy distribution of the radiation as a function of frequency by multiplying both sides of (3.99) by the energy carried by each photon:

$$\frac{dU(\nu)}{d\nu} = \frac{8\pi h}{c^3} L^3 \frac{\nu^3}{e^{\frac{h\nu}{kT}} - 1}. \tag{3.100}$$

This distribution was deduced for the first time by Planck and was indeed named after him.

It is evident that at small frequencies Planck distribution is practically equal to that in (3.96). Instead at high frequencies energy quantization leads to an exponential cut in the energy distribution which eliminates the paradox of an infinite internal energy and of an infinite specific heat. We notice that the phenomenon suppressing the high energy modes in the computation of the specific heat is the same leading to a vanishing specific heat for the harmonic oscillator when $kT \ll h\nu$ (the system cannot absorb a quantity of energy less the minimal quantum $h\nu$) and indeed, as we have already stressed at the end of Section 2.7, the radiation field in a box can be considered as an infinite collection of independent harmonic oscillators of frequencies given by the resonant frequencies of the box.

Suggestions for Supplementary Readings

- F. Reif: *Statistical Physics - Berkeley Physics Course*, volume 5 (Mcgraw-Hill Book Company, New York 1965)
- E. Schrödinger: *Statistical Thermodynamics* (Cambridge University Press, Cambridge 1957)
- T. L. Hill: *An Introduction to Statistical Thermodynamics* (Addison-Wesley Publishing Company Inc., Reading 1960)
- F. Reif: *Fundamentals of Statistical and Thermal Physics* (Mcgraw-Hill Book Company, New York 1965)

Problems

3.1. We have to place four distinct objects into 3 boxes. How many possible different distributions can we choose? What is the multiplicity $\mathcal{M}$ of each distribution? And its probability p?

Answer: We can make 3 different choices for each object, therefore the total number of possible choices is $3^4 = 81$. The total number of possible distributions is instead given by all the possible choices of non-negative integers n_1, n_2, n_3 with $n_1 + n_2 + n_3 = 4$, i.e. $(4+1)(4+2)/2 = 15$. There are in particular 3 distributions like $(4, 0, 0)$, each with $p = \mathcal{M}/81 = 1/81$; 6 like $(3, 1, 0)$, with $p = \mathcal{M}/81 = 4/81$; 3 like $(2, 2, 0)$, with $p = \mathcal{M}/81 = 6/81$; 3 like $(2, 1, 1)$, with $p = \mathcal{M}/81 = 12/81$.

3.2. The integer number k can take values in the range between 0 and 8 according to the binomial distribution:

$$P(k) = \frac{1}{2^8} \binom{8}{k}.$$

Compute the mean value of k and its mean quadratic deviation.

Answer: $\bar{k} = 4$; $\langle (k - \bar{k})^2 \rangle = 2$.

3.3. Let us consider a system which can be found in 4 possible states, enumerated by the index $k = 0, 1, 2, 3$ and with energy $E_k = \epsilon k$, where $\epsilon = 10^{-2}$ eV. The system is at thermal equilibrium at room temperature $T \simeq 300°$K. What is the probability of the system being in the highest energy state?

Answer: $Z = \sum_{k=0}^{3} e^{-\beta \epsilon k}$; $U = (1/Z) \sum_{k=0}^{3} \epsilon k e^{-\beta \epsilon k} \simeq 1.035 \ 10^{-2}$ eV; $P_{k=3} = (1/Z) e^{-3\beta \epsilon} \simeq 0.127$.

3.4. A biatomic molecule is made up of two particles of equal mass $M = 10^{-27}$ Kg which are kept at a fixed distance $L = 4 \ 10^{-10}$ m. A set of $N = 10^9$ such systems, which are not interacting among themselves, is in thermal equilibrium at a temperature $T = 1°$K. Estimate the number of systems which have a non-vanishing angular momentum (computed with respect to their center of mass), i.e. the number of rotating molecules, making use of the fact that the number of states with angular momentum $n\hbar$ is equal to $2n + 1$.

Answer: If we quantize rotational energy according to Bohr, then the possible energy levels are $E_n = n^2 \hbar^2 / (ML^2) \simeq 4.3 \ 10^{-4} n^2$ eV, each corresponding to $2n+1$ different states. These states are occupied according to the Canonical Distribution. The partition function is $Z = \sum_{n=0}^{\infty} (2n+1) e^{-E_n/kT}$. If $T = 1°$K, then $kT \simeq 0.862 \ 10^{-4}$ eV, hence $e^{-E_n/kT} \simeq e^{-5.04 n^2} \simeq (6.47 \ 10^{-3})^{n^2}$. Therefore the two terms with $n = 0, 1$ give a very good approximation of the partition function, $Z \simeq 1 + 1.94 \ 10^{-2}$. The probability that a molecule has $n = 0$ is $1/Z$, hence the number of rotating molecules is $N_R = N(1 - 1/Z) \simeq 1.9 \; 10^7$. If we instead make use of Sommerfeld's perfected theory, implying $n^2 \to n(n+1)$ in the expression for E_n, we obtain $Z \simeq 1 + 1.25 \; 10^{-4}$ and $N_R = N(1 - 1/Z) \simeq 1.25 \; 10^5$. In the present situation, being quantum effects quite relevant, the use of the Sommerfeld's correct formula for angular momentum quantization in place of simple Bohr's rule makes a great difference.

3.5. Consider again Problem 3.4 in case the molecules are in equilibrium at room temperature, $T \simeq 300°$K. Compute also the average energy of each molecule.

Answer: In this case the partition function is, according to Sommerfeld's theory:

$$Z = \sum_{n=0}^{\infty} (2n + 1)e^{-\alpha n(n+1)} ,$$

with $\alpha \simeq 0.0168$. Since $\alpha \ll 1$, $(2n + 1)e^{-\alpha n^2}$ is the product of a linear term times a slowly varying function of n, hence the sum can be replaced by an integral

$$Z \simeq \int_0^\infty dn (2n + 1)e^{-\alpha n(n+1)} = \frac{1}{\alpha} \simeq 60$$

hence the number of non-rotating molecules is $N_{NR} = N/Z \simeq 1.67 \; 10^7$. From $Z \simeq 1/\alpha = kTML^2/\hbar^2$, we get $U = -\partial/\partial\beta \ln Z = kT$, in agreement with equipartition of energy.

3.6. A system in thermal equilibrium admits 4 possible states: the fundamental state having zero energy plus three degenerate excited states of energy ϵ. Discuss the dependence of its mean energy on the temperature T.

Answer:

$$U = \frac{3\epsilon \; e^{-\epsilon/kT}}{1 + 3e^{-\epsilon/kT}} \; ; \qquad \lim_{T \to 0} U(T) = 0 \; ; \qquad \lim_{T \to \infty} U(T) = \frac{3}{4}\epsilon \; .$$

3.7. A simple pendulum of length $l = 10$ cm and mass $m = 10$ g is placed on Earth's surface in thermal equilibrium at room temperature, $T = 300°$K. What is the mean quadratic displacement of the pendulum from its equilibrium point?

Answer: The potential energy of the pendulum is, for small displacements $s \ll l$, $mgs^2/2l$. From the energy equipartition theorem we infer $mgs^2/2l \simeq kT/2$, hence $\sqrt{\langle s^2 \rangle} \simeq \sqrt{kTl/mg} = 2 \; 10^{-10}$ m .

3.8. A massless particle is constrained to move along a segment of length L; therefore its wave function vanishes at the ends of the segment. The system is in equilibrium at a temperature T. Compute its mean energy as well as the specific heat at fixed L. What is the force exerted by the particle on the ends of the segment?

Answer: Energy lavels are given by $E_n = nc\pi\hbar/L$, hence $Z = \sum_{n=1}^{\infty} e^{-\beta E_n} = 1/(e^{\beta c\pi\hbar/L} - 1)$ from which the mean energy follows

$$U = -\frac{\partial \ln Z}{\partial \beta} = \frac{c\pi\hbar}{L} \frac{1}{(1 - e^{-c\pi\hbar/LkT})} ,$$

and the specific heat

$$C_L = k(c\pi\hbar/LkT)^2 \frac{e^{-c\pi\hbar/LkT}}{(1 - e^{-c\pi\hbar/LkT})^2} ,$$

which vanishes at low temperatures and approaches k at high temperatures; notice that the equipartition principle does not hold in its usual form in this example, since the energy is not quadratic in the momentum, hence we have k instead of $k/2$. The equation of state can be obtained making use of (3.29) and (3.35), giving for the force $F = (1/\beta)(\partial \ln Z/\partial L) = U/L$. Hence at high temperatures we have $FL = kT$.

3.9. Consider a system made up of N distinguishable and non-interacting particles which can be found each in two possible states of energy 0 and ϵ. The system is in thermal equilibrium at a temperature T. Compute the mean energy and the specific heat of the system.

Answer: The partition function for a single particle is $Z_1 = 1 + e^{-\epsilon/kT}$. That for N independent particles is $Z_N = Z_1^N$. Therefore the average energy is

$$U = \frac{N\epsilon}{1 + e^{\epsilon/kT}}$$

and the specific heat is

$$C = Nk \left(\frac{\epsilon}{kT}\right)^2 \frac{e^{\epsilon/kT}}{(e^{\epsilon/kT} + 1)^2} .$$

3.10. A system consists of a particle of mass m moving in a one dimensional potential which is harmonic for $x > 0$ ($V = kx^2/2$) and infinite for $x < 0$. If the system as at thermal equilibrium at a temperature T, compute its average energy and its specific heat.

Answer: The wave function must vanish in the origin, hence the possible energy levels are those of the harmonic oscillator having an odd wave function. In particular, setting $\omega = \sqrt{k/m}$, we have $E_n = (2n + 3/2)\hbar\omega$, with $n = 0, 1, ...$ The partition function is

$$Z = \frac{e^{-3\beta\hbar\omega/2}}{1 - e^{-2\beta\hbar\omega}} ,$$

so that the average energy is

$$U = \frac{3\hbar\omega}{2} + \frac{2\hbar\omega}{e^{2\beta\hbar\omega} - 1}$$

and the specific heat is

$$C = k(2\beta\hbar\omega)^2 \frac{e^{2\beta\hbar\omega}}{(e^{2\beta\hbar\omega} - 1)^2} .$$

3.11. Compute the average energy of a classical three-dimensional isotropic harmonic oscillator of mass m and oscillation frequency $\nu = 2\pi\omega$ in equilibrium at temperature T.

Answer: The state of the classical system is assigned in terms of the momentum p and the coordinate x of the oscillator, it is therefore represented by a point in phase space corresponding to an energy $E(p,x) = p^2/2m + m\omega^2 x^2/2$. The canonical partition function can therefore be written as an integral over phase space

$$Z = \int \frac{d^3p\, d^3x}{\Delta} e^{-\beta p^2/2m} e^{-\beta m\omega^2 x^2/2}$$

where Δ is an arbitrary effective volume in phase space needed to fix how we count states (that is actually not arbitrary according to the quantum theory, which requires $\Delta \sim h^3$). A simple computation of Gaussian integrals gives $Z = \Delta^{-1}(2\pi/\omega\beta)^3$, hence $U = -(\partial/\partial\beta)\ln Z = 3kT$, in agreement with equipartition of energy.

3.12. A particle of mass m moves in the $x-y$ plane under the influence of an anisotropic harmonic potential $V(x,y) = m(\omega_x^2 x^2/2 + \omega_y^2 y^2/2)$, with $\omega_y \ll \omega_x$. Therefore the energy levels coincide with those of a system made up of two distinct particles moving in two different one dimensional harmonic potentials corresponding respectively to ω_x and ω_y. The system is in thermal equilibrium at a temperature T. Compute the specific heat and discuss its behaviour as a function of T.

Answer: The partition function is the product of the partition functions of the two distinct harmonic oscillators, hence the average energy and the specific heat will be the sum of the respective quantities. In particular

$$C = \frac{(\hbar\omega_x)^2}{kT^2} \frac{e^{\beta\hbar\omega_x}}{(e^{\beta\hbar\omega_x}-1)^2} + \frac{(\hbar\omega_y)^2}{kT^2} \frac{e^{\beta\hbar\omega_y}}{(e^{\beta\hbar\omega_y}-1)^2} \; .$$

We have three different regimes: $C \sim 0$ if $kT \ll \hbar\omega_y$, $C \sim k$ if $\hbar\omega_y \ll kT \ll \hbar\omega_x$ and finally $C \sim 2k$ if $kT \gg \hbar\omega_x$.

3.13. Consider a biatomic gas, whose molecules can be described schematically as a pair of pointlike particles of mass $M = 10^{-27}$ Kg, which are kept at an equilibrium distance $d = 2\ 10^{-10}$ m by an elastic force of constant $K = 11.25$ N/m. A quantity equal to 1.66 gram atoms of such gas is contained in volume $V = 1$ m^3. Discuss the qualitative behaviour of the specific heat of the system as a function of temperature. Consider the molecules as non-interacting and as if each were contained in a cubic box with reflecting walls of size $L^3 = V/N$, where N is the total number of molecules.

Answer: Three different energy scales must be considered. The effective volume available for each molecule sets an energy scale $E_1 = \hbar^2\pi^2/(4ML^2) \simeq 1.7\ 10^{-6}$ eV, which is equal to the fundamental level for a particle of mass $2M$ in a cubic box, corresponding to a temperature $T_1 = E_1/k \simeq 0.02°$K. The minimum rotational energy is instead, according to Sommerfeld, $E_2 = \hbar^2/(Md^2) \simeq 3.5\ 10^{-3}$ eV, corresponding to a temperature $T_2 = E_2/k \simeq 40°$K. Finally, the fundamental oscillation energy is $E_3 = \hbar\sqrt{2K/M} \simeq 0.098$ eV, corresponding to a temperature $T_3 = E_2/k \simeq 1140°$K. For $T_1 \ll T \ll T_2$ the system can be described as a classical perfect gas of pointlike particles, since rotational and vibrational modes are not yet excited, hence the specific heat per molecule is $C \sim 3k/2$. For $T_2 \ll T \ll T_3$ the system can be described as a classical perfect gas of rigid rotators, hence $C \sim 5k/2$. Finally, for $T \gg T_3$ also the (one-dimensional) vibrational mode is excited and $C \sim 7k/2$. This roughly reproduces, from a qualitative point of view and with an appropriate rescaling of parameters, the observed behavior of real biatomic gasses.

3.14. Consider a system made up of two identical fermionic particles which can occupy 4 different states. Enumerate all the possible choices for the occupation numbers of the single particle states. Assuming that the 4 states have the following energies: $E_1 = E_2 = 0$ and $E_3 = E_4 = \epsilon$ and that the system is in thermal equilibrium at a temperature T, compute the mean occupation number of one of the first two states as a function of temperature.

Answer: There are six different possible states for the whole system characterized by the following occupation numbers (n_1, n_2, n_3, n_4) for the single particle states: $(1,1,0,0)$, $(1,0,1,0)$, $(1,0,0,1)$, $(0,1,1,0)$, $(0,1,0,1)$, $(0,0,1,1)$. The corresponding energies are $0, \epsilon, \epsilon, \epsilon, \epsilon, 2\epsilon$. The mean occupation number of the first single particle state (i.e. $\langle n_1 \rangle$) is then given by averaging the value of n_1 over the 6 possible states weighted using the Canonical Distribution, i.e.

$$\langle n_1 \rangle = \frac{(1 + 2e^{-\beta\epsilon})}{(1 + 4e^{-\beta\epsilon} + e^{-2\beta\epsilon})} .$$

3.15. A system, characterized by 3 different single particle states, is filled with 4 identical bosons. Enumerate the possible states of the system specifying the corresponding occupation numbers. Discuss also the case of 4 identical fermions.

Answer: The possible states can be enumerated by indicating all possible choices for the occupation numbers n_1, n_2, n_3 satisfying $n_1 + n_2 + n_3 = 4$. That leads to 15 different states. In the case of fermions, since $n_i = 0, 1$, the constraint on the total number of particles cannot be satisfied and there is actually no possible state of the system.

3.16. A system, characterized by two single particle states of energy $E_1 = 0$ and $E_2 = \epsilon$, is filled with 4 identical bosons. Enumerate all possible choices for the occupation numbers. Assuming that the system is in thermal contact with a reservoir at temperature T and that $e^{-\beta\epsilon} = \frac{1}{2}$, compute the probability of all particles being in the fundamental state. Compare the answer with that for distinguishable particles.

Answer: Since the occupation numbers must satisfy $N_1 + N_2 = 4$, the possible states are identified by the value of, for instance, N_2, in the case of bosons ($N_2 = 0, 1, 2, 3, 4$), and have energy ϵN_2. In the case of distinguishable particles there are instead $4!/(N_2!(4 - N_2)!)$ different states for each value of N_2. The probability of all the particles being in the fundamental state is $16/31$ in the first case and $(2/3)^4$ in the second case: notice that this probability is highly enhanced in the case of bosons.

3.17. Consider a gas of electrons at zero temperature. What is density at which relativistic effects show up? Specify the answer by finding the density for which electrons occupy states corresponding to velocities $v = \sqrt{3}\,c/2$.

Answer: At $T = 0$ electrons occupy all levels below the Fermi energy E_F, or equivalently below the corresponding Fermi momentum p_F. To answer the question we must impose that

$$p_F = \frac{m_e v}{\sqrt{1 - v^2/c^2}} = \sqrt{3}\, m_e c \ .$$

On the other hand, the number of states below the Fermi momentum, assuming the gas is contained in a cubic box of size L, is

$$N = \frac{p_F^3 L^3}{3\hbar^3 \pi^2} \, ,$$

hence $\rho = p_F^3/(3\pi^2 \hbar^3) \simeq 3.04 \ 10^{36}$ particles/m^3.

3.18. The density of states as a function of energy in the case of free electrons is given in (3.64). However in a conduction band the distribution may have a different dependence on energy. Let us consider for instance the simple case in which the density is constant, $dn_E/dE = \gamma\, V$, where $\gamma = 8\ 10^{47}$ m^{-3} J^{-1}, the energy varies from zero to $E_0 = 1$ eV and the electronic density is $\rho \equiv \bar{n}/V = 6\ 10^{28}$ m^{-3}. For T not much greater than room temperature it is possible to assume that the bands above the conduction one are completely free, while those below are completely occupied, hence the thermal properties can be studied solely on the basis of its conduction band. Under these assumptions, compute how the chemical potential μ depends on temperature.

Answer: The average total number of particles comes out to be $\bar{N} = \int_0^{E_0} n(\epsilon)g(\epsilon)d\epsilon$. The density of levels is $g(\epsilon) = dn_E/dE = \gamma\, V$ and the average occupation number is $n(\epsilon) = 1/(e^{\beta(\epsilon - \mu)} + 1)$. After computing the integral and solving for μ we obtain

$$\mu = kT \ln\left(\frac{e^{\rho/(\gamma kT)} - 1}{1 - e^{\rho/(\gamma kT)}e^{-E_0/kT}} \right) \ .$$

It can be verified that, since by assumption $\rho/\gamma < E_0$, in the limit $T \to 0$ μ is equal to the Fermi energy $E_F = \rho/\gamma$. Instead, in the opposite large temperature limit, $\mu \to -kT \ln(1 - \gamma E_0/\rho)$, hence the distribution of electrons in energy would be constant over the band and simply given by $n(\epsilon)g(\epsilon) = \rho V/E_0$, but of course in this limit we cannot neglect te presence of other bands. Notice also that in this case, due to the different distribution of levels in energy, we have $\mu \to +\infty$ instead of $\mu \to -\infty$ as $T \to \infty$.

3.19. Compute the mean quadratic velocity for a rarefied and ideal gas of particles of mass $M = 10^{-20}$ Kg in equilibrium at room temperature.

Answer: According to Maxwell distribution, $\langle v^2 \rangle = 3kT/m \simeq 1.2$ m^2/s^2.

3.20. The modern theory of cosmogenesis suggests that cosmic space contain about 10^8 neutrinos per cubic meter and for each species of these particles. Neutrinos can be considered, in a first approximation, as massless fermion having a single spin state instead of two, as for electrons; they belong to 6 different species. Assuming that each species be independent of the others, compute the corresponding Fermi energy.

Answer: Considering a gas of neutrinos placed in a cubic box of size L, the number of single particle states with energy below the Fermi energy E_F is given, for massless particles, by $N_{E_F} = (\pi/6)E_F^3 L^3/(\pi\hbar c)^3$. Putting that equal to the average number of particles in the box, $\bar{N} = \rho L^3$, we have $E_F = \hbar c \, (6\pi^2\rho)^{1/3} \simeq 3.38 \cdot 10^{-4}$ eV. Notice that, of course, the finite size of the box disappears in the final result as a consequence of the continuum approximation for the level distribution in energy, which is better the larger the volume is.

3.21. Suppose now that neutrinos must be described as particles of mass $m_\nu \neq 0$. Consider again Problem 3.20 and give the exact relativistic formula expressing the Fermi energy in terms of the gas density ρ.

Answer: The formula expressing the total number of particles $N = L^3\rho$ in terms of the Fermi momentum p_F is:

$$N_{E_F} = (\pi/6)p_F^3 L^3/(\pi\hbar)^3 ,$$

hence

$$E_F = \sqrt{m_\nu^2 c^4 + p_F^2 c^2} = \sqrt{m_\nu^2 c^4 + (6\pi^2\rho)^{2/3}(\hbar c)^2} .$$

3.22. Compute the internal energy and the pressure at zero temperature for the system described in Problem 3.20, i.e. for a gas of massless fermions with a single spin state and a density $\rho \equiv \bar{n}V = 10^8$ m^{-3}.

Answer: The density of internal energy is

$$U/V = (81\pi^2\rho^4/32)^{\frac{1}{3}}\hbar c \simeq 4.29 \; 10^{-15} \; \text{J/m}^3 ,$$

and the pressure

$$P = U/3V \simeq 1.43 \; 10^{-15} \; \text{Pa} .$$

Notice that last result is different from what obtained for electrons, equation (3.79): the factor $1/3$ in place of $2/3$ is a direct consequence of the linear dependence of energy on momentum taking place for massless or ultrarelativistic particles, in contrast with the quadratic behavior which is valid for (massive) non-relativistic particles.

3.23. 10^3 bosons move in a harmonic potential corresponding to a frequency ν such that $h\nu = 1$ eV. Considering that the mean occupation number of the m-th level of the oscillator is given by the Bose–Einstein distribution: $n_m = \left(e^{\beta(hm\nu-\mu)} - 1\right)^{-1}$, compute the chemical potential assuming $T = 300°$K.

Answer: The total number of particles, $N = 1000$, can be written as

$$N = \sum_m n_m = \frac{1}{e^{-\beta\mu} - 1} + \frac{1}{Ke^{-\beta\mu} - 1} + \frac{1}{K^2 e^{-\beta\mu} - 1} + \cdots$$

where $K = \exp(h\nu/kT) \simeq e^{40}$. Since K is very large and $\exp(-\beta\mu) > 1$ ($\mu \le 0$ for bosons), it is clear that only the first term is appreciably different from zero. Hence

$$\exp(-\beta\mu) = 1 + \frac{1}{N}$$

and finally $\mu \simeq -2.5 \ 10^{-5}$ eV.

3.24. A system is made up of N identical bosonic particles of mass m moving in a one-dimensional harmonic potential $V(x) = m\omega^2 x^2/2$. What is the distribution of occupation numbers corresponding to the fundamental state of the system? And that corresponding to the first excited state? Determine the energy of both states.

If $\hbar\omega = 0.1$ eV and if the system is in thermal equilibrium at room temperature, $T = 300°$K, what is the ratio R of the probability of the system being in the first excited state to that of being in the fundamental one? How the last answer changes in case of distinguishable particles?

Answer: In the fundamental state all particles occupy the single particle state of lowest energy $\hbar\omega/2$, hence $E = N\hbar\omega/2$, while in the first excited state one of the N particles has energy $3\hbar\omega/2$, hence $E = (N + 2)\hbar\omega/2$, The fundamental state has degeneracy 1 both for identical and distinguishable particles. The first excited state has denegeracy 1 in the case of bosons while the degeneracy is N in the other case, since it makes sense to ask which of the N particles has energy $3\hbar\omega/2$. Therefore $R = e^{-\hbar\omega/kT} \simeq 0.021$ for bosons and $R = N \ 0.021$ in the second case. For large N the probability of the system being excited is much suppressed in the case of bosons with respect to the case of distinguishable particles.

3.25. Consider again Problem 3.24 in the case of fermions having a single spin state and for $\hbar\omega = 1$ eV and $T \simeq 1000°$K.

Answer: In the fundamental state of the system the first N levels of the harmonic oscillator are occupied, hence its energy is $E_0 = \sum_{i=0}^{N-1}(n + 1/2)\hbar\omega = (N^2/2)\hbar\omega$. The minimum possible excitation of this state corresponds to moving the fermion of highest energy up to the next free level, hence the energy of the first excited state is $E_1 = E_0 + \hbar\omega$. The ratio R is equal to $e^{-\hbar\omega/kT} = 9.12 \ 10^{-6}$.

3.26. A system is made up of $N = 10^8$ electrons which are free to move along a conducting cable of length $L = 1$ cm, which can be roughly described as a one dimensional segment with reflecting endpoints. Compute the Fermi energy of the system, taking also into account the spin degree of freedom.

Answer: $E_F = \hbar^2 N^2 \pi^2 / (8mL^2) = 1.5 \ 10^{-18}$ J.

A

Quadrivectors

A synthetic description of Lorentz transformations and of their action on physical observables can be given in terms of matrix algebra. The coordinates of a space-time event in the reference frame O are identified with the elements of a column matrix ξ:

$$\xi \equiv \begin{pmatrix} ct \\ x \\ y \\ z \end{pmatrix},$$

(A.1)

while Lorentz transformations from O to a new reference frame O', which are given in (1.13) and (1.14), are associated with a matrix

$$\bar{\Lambda} \equiv \begin{pmatrix} \frac{1}{\sqrt{1-\frac{v^2}{c^2}}} & 0 & 0 & \frac{-v}{c\sqrt{1-\frac{v^2}{c^2}}} \\ 0 & 1 & 0 & 0 \\ 0 & 0 & 1 & 0 \\ \frac{-v}{c\sqrt{1-\frac{v^2}{c^2}}} & 0 & 0 & \frac{1}{\sqrt{1-\frac{v^2}{c^2}}} \end{pmatrix}$$

(A.2)

defined so that the column matrix corresponding to the coordinates of the same space-time event in frame O' is given by:

$$\xi' \equiv \begin{pmatrix} ct' \\ x' \\ y' \\ z' \end{pmatrix} = \bar{\Lambda}\xi \equiv \begin{pmatrix} \frac{1}{\sqrt{1-\frac{v^2}{c^2}}} & 0 & 0 & \frac{-v}{c\sqrt{1-\frac{v^2}{c^2}}} \\ 0 & 1 & 0 & 0 \\ 0 & 0 & 1 & 0 \\ \frac{-v}{c\sqrt{1-\frac{v^2}{c^2}}} & 0 & 0 & \frac{1}{\sqrt{1-\frac{v^2}{c^2}}} \end{pmatrix} \begin{pmatrix} ct \\ x \\ y \\ z \end{pmatrix}.$$

(A.3)

The products above are intended to be row by column products. To be more specific, indicating by $a_{i,j}$, with $i = 1, \ldots, M$ and $j = 1, \ldots, N$, the element corresponding to the i-th row and the j-th column of the $M \times N$ matrix A (i.e. A has M rows and N columns), and by $b_{l,m}$ the element of the i-th row and the j-th column of the $N \times P$ matrix B, the product AB is a $M \times P$ matrix whose generic element is given by:

$$(AB)_{i,m} = \sum_{j=1}^{N} a_{i,j} b_{j,m} \,. \tag{A.4}$$

As we have discussed in Chapter 2, a generic Lorentz transformation corresponds to a homogeneous linear transformations of the event coordinates which leaves invariant the following quadratic form:

$$\xi^2 \equiv x^2 + y^2 + z^2 - c^2 t^2 \,; \tag{A.5}$$

moreover, the direction of time (time arrow) is the same in both frames and if the spatial Cartesian coordinates in O are right(left)-handed, they will be right(left)-handed also in O'. In matrix algebra the quadratic form can be easily built up by introducing the *metric*

$$g \equiv \begin{pmatrix} -1 & 0 & 0 & 0 \\ 0 & 1 & 0 & 0 \\ 0 & 0 & 1 & 0 \\ 0 & 0 & 0 & 1 \end{pmatrix}, \tag{A.6}$$

and defining:

$$\xi^2 = \xi^T g \, \xi \equiv \begin{pmatrix} ct' & x' & y' & z' \end{pmatrix} \begin{pmatrix} -1 & 0 & 0 & 0 \\ 0 & 1 & 0 & 0 \\ 0 & 0 & 1 & 0 \\ 0 & 0 & 0 & 1 \end{pmatrix} \begin{pmatrix} ct \\ x \\ y \\ z \end{pmatrix}, \tag{A.7}$$

where the products are still intended to be row by column.

The symbol T used in the left-hand side stands for transposition, i.e. the operation corresponding to exchanging rows with columns. Notice that, maintaining the row by column convention for matrix product, the transpose of the product of two matrices is the inverted product of the transposes:

$$(AB)^T = B^T A^T \,. \tag{A.8}$$

The condition that Lorentz transformations leave the quadratic form in (A.7) invariant can be now expressed as a simple matrix equation for the generic transformation Λ:

$$\xi'^T g \, \xi' = \xi^T \Lambda^T g \Lambda \xi \equiv \xi^T g \, \xi \qquad \forall \, \xi \qquad \Longrightarrow \Lambda^T g \Lambda = g \,. \tag{A.9}$$

The additional requirements are that the determinant of Λ be one and that its first diagonal element (i.e. first row, first column) be positive.

Equation (A.9) can be put in a different form by making use of the evident relation $g^2 = I$, where I is the identity matrix (its element (i, j) is equal to 1 if $i = j$, to 0 if $i \neq j$ and of course $AI = IA = A$ for every matrix A). Multiplying both sides of last equation in (A.9) by g on the left, we obtain:

$$g \Lambda^T g \Lambda = I \Longrightarrow g \Lambda^T g = \Lambda^{-1} \,, \tag{A.10}$$

where the inverse matrix A^{-1} of a generic matrix A is defined by the equation $A^{-1}A = AA^{-1} = I$.

Column vectors in relativity are called quadrivectors. Some specifications about the action of Lorentz transformations over them are however still needed. Let us consider a function $f(\xi)$ depending on the space-time event ξ, which we suppose to have continuous derivatives and to be Lorentz invariant, i.e. such that $f(\Lambda\xi) = f(\xi)$. The partial derivatives of f with respect to the coordinates of the event form a column vector:

$$\partial f(\xi) \equiv \begin{pmatrix} \frac{\partial f(\xi)}{\partial ct} \\ \frac{\partial f(\xi)}{\partial x} \\ \frac{\partial f(\xi)}{\partial y} \\ \frac{\partial f(\xi)}{\partial z} \end{pmatrix}. \tag{A.11}$$

However this vector does not transform like ξ when changing reference frame. Indeed in O' we have:

$$\partial' f(\xi') \equiv \begin{pmatrix} \frac{\partial f(\xi')}{\partial ct'} \\ \frac{\partial f(\xi')}{\partial x'} \\ \frac{\partial f(\xi')}{\partial y'} \\ \frac{\partial f(\xi')}{\partial z'} \end{pmatrix} = \begin{pmatrix} \frac{\partial f(\xi)}{\partial ct'} \\ \frac{\partial f(\xi)}{\partial x'} \\ \frac{\partial f(\xi)}{\partial y'} \\ \frac{\partial f(\xi)}{\partial z'} \end{pmatrix}. \tag{A.12}$$

To proceed further it is necessary to make use of (A.10), which gives $\xi = g\Lambda^T g\xi'$, and to apply the chain rule for partial derivatives. This states that, if the variables q_i, $i = 1, \ldots, n$ depend on the variables q_i' through the function $q_i = Q_i(q')$, and vice versa $q_i' = \tilde{Q}_i(q)$, then for the partial derivatives of a function $F(q)$ we have:

$$\frac{\partial F(q)}{\partial q_i'} = \sum_{l=1}^{n} \frac{\partial Q_l(\tilde{Q}(q))}{\partial q_i'} \frac{\partial F(q)}{\partial q_l}. \tag{A.13}$$

Interpreting last expression as a matrix product, it is clear that i is a row index while l is a column one. In the specific case of (A.12), $Q(q')$ is a linear function corresponding to $g\Lambda^T g \; \xi'$, while $\partial Q_l(\tilde{Q}(q))/\partial q_i'$ corresponds to the matrix $(g\Lambda^T g)^T = g\Lambda g$. Indeed, as noticed above and as follows from the definition of matrix product, index i refers to the row of the first matrix and to the column of the second.

We have therefore the matrix relation:

$$\partial' f = g\Lambda g \partial f \longrightarrow g\partial' f = \Lambda g \partial f, \tag{A.14}$$

showing that it is the product of g by the vector of partial derivatives which transforms like the coordinates of a space-time event, rather than the partial derivatives themselves. Hence, if we want to call *quadrivector* a column vector transforming like coordinates, $g\partial f$ is a quadrivector while ∂f is not.

In mathematical language the vector of coordinates is usually called a *contravariant* quadrivector, while that of partial derivatives is called a *covariant* quadrivector; the difference is specified by a different position of indices, which go up in the contravariant case. Notice however that this distinction is important only in the case of non-linear coordinate transformations, like in the case of curved manifolds. In our case the introduction of the two kind of vectors is surely not worthwhile.

Indices may be convenient in the case of quantities transforming like the tensorial product of more than one vector, i.e. like the products of different components of quadrivectors: that is the case, for instance, of the electric and magnetic fields, but not of the vector potential.

Having verified that several physical quantities transform like quadrivectors, let us notice that an invariant, i.e. a quantity which is the same for all inertial observers, can be associated with each pair of quadrivectors η and ζ:

$$\eta^T g\, \zeta = \zeta^T g\, \eta \;\Rightarrow\; \eta'^T g\, \zeta' = (\Lambda\eta)^T g\, \Lambda\zeta = \eta^T \Lambda^T g\, \Lambda\zeta = \eta^T g\, \zeta\,. \quad (A.15)$$

In analogy with rotations, this invariant is usually called scalar product and indicated by $\eta \cdot \zeta$. A quite relevant example of scalar product is represented by de Broglie's phase, which is given by $-\epsilon \cdot \xi/\hbar = (\boldsymbol{p} \cdot \boldsymbol{r} - Et)/\hbar$, i.e. it is the scalar product of the coordinate quadrivector ξ and of the energy-momentum quadrivector ϵ of a given particle.

B

The Schrödinger Equation in a Central Potential

In the case of a particle moving in three dimensions under the influence of a central force field, the symmetry properties of the problem play a dominant role. The problem is that of solving the stationary Schrödinger equation:

$$-\frac{\hbar^2}{2m}\nabla^2\psi_E(\boldsymbol{r}) + V(r)\psi_E(\boldsymbol{r}) = E\psi_E(\boldsymbol{r}) , \tag{B.1}$$

where the position of the particle has been indicated by the three-vector $\boldsymbol{r}$, i.e. by the column matrix with components x, y, z, in the same language of Appendix A. The symmetries of the problem correspond to all possible rotations around the origin, identified with 3×3 orthogonal matrices R, hence satisfying $R^T = R^{-1}$, and acting on the position vectors as follows: $\boldsymbol{r} \rightarrow R\boldsymbol{r}$. The invariance of the potential under rotations implies that the rotated wave function $\psi_E(R^{-1}\boldsymbol{r})$ is also a solution of (B.1) corresponding to the same energy.

We can consider, among all possible rotations, those around one particular axis, for instance the z axis, which transform $x \rightarrow x' = x\cos\phi - y\sin\phi$ and $y \rightarrow y' = y\cos\phi + x\sin\phi$, while z is left unchanged. An equivalent way of writing this rotation, making use of complex combinations of coordinates, is the following: $x'_{\pm} \equiv x' \pm iy' = e^{\pm i\phi}x_{\pm}$, and $z' = z$. In spherical coordinates (r, θ, φ), defined by $x = r\sin\theta\cos\varphi$, $y = r\sin\theta\sin\varphi$, and $z = r\cos\theta$, the same rotation is equivalent to the translation $\varphi \rightarrow \varphi' = \varphi + \phi$. The symmetry principle we have often referred to in this text affirms that we can choose the solutions of (B.1) such that they do not change but for a phase $\Phi(\phi)$, $\psi_E \longrightarrow e^{i\Phi}\psi_E$ under the particular rotation above. This phase must necessarily be a linear function of ϕ, as it is clear by observing that, if we consider two subsequent rotations around the same axis with angles ϕ and ϕ', we have $\Phi(\phi) + \Phi(\phi') = \Phi(\phi + \phi')$. Combining this result with the condition that, if $\phi = 2\pi$, the wave function must be left unchanged, i.e. $\Phi(2\pi) = 2\pi m$ with m any relative integer, we obtain, in spherical coordinates:

$$\psi_{E,m}(\boldsymbol{r}) \equiv \psi_{E,m}(r, \theta, \varphi) = \hat{\psi}_{E,m}(r, \theta)e^{im\varphi} . \tag{B.2}$$

It is an easy exercise to verify that

$$\mathrm{i}\hbar(y\partial_x - x\partial_y)\psi_{E,m}(\boldsymbol{r}) = -\mathrm{i}\hbar\frac{\partial}{\partial\varphi}\psi_{E,m}(\boldsymbol{r}) = m\hbar\psi_{E,m}(\boldsymbol{r}), \qquad (\mathrm{B.3})$$

thus showing that the wave function satisfies Bohr's quantization rule for the z component of angular momentum.

Notice the evident benefit deriving from making use of vectors with complex components $(x_{\pm} , z)$ to indicate the position. In the same language of Appendix A, these are column matrices where the first element (x_+) is the complex conjugate of the second (x_-), while the third (z) is real. Indicating as usual by $\boldsymbol{v}$ these column matrices and again by R the matrices corresponding to rotations, so that $\boldsymbol{v}' = R\boldsymbol{v}$, we can easily find the constraints satisfied by R in the new complex notation. The squared length of a vector is $v^2 \equiv \boldsymbol{v}^T g \boldsymbol{v}$, where the only non-vanishing components of the metric matrix $g_{i,j}$ are $g_{+,-} = g_{-,+} = \frac{1}{2}$ and $g_{3,3} = 1$ (indeed we have $r^2 = x_+x_- + z^2$), so that the condition of length invariance for vectors can be written as $R^T g R = g$, or equivalently $g^{-1}R^T g R = I$, if I is the identity matrix and $g^{-1}g = gg^{-1} = I$. The matrix g^{-1} satisfies $Rg^{-1}R^T = g^{-1}$ and its non-vanishing components are $g_{+,-}^{-1} = g_{-,+}^{-1} = 2$ and $g_{3,3}^{-1} = 1$.

Let us now introduce a generic symmetric tensor of rank n, which is a quantity with n indices, $T_{i_1,..,i_n}$, transforming under rotations according to $T'_{i_1,..,i_n} = \sum_{j_1,..,j_n} R_{i_1 j_1} \cdots R_{i_n j_n} T_{j_1,..,j_n}$. In particular, a tensor of rank 2 is a square matrix T such that $T' = RTR^T$. According to the considerations above, the quantity g^{-1} can be considered as an invariant tensor. In the case of rank 2 we can also introduce the quantity $\sum_{i,j} T_{ij}g_{ij} \equiv Tr(T)$ which is called the trace of the tensor and is invariant under rotations. Since g^{-1} is invariant as well, it follows that the new tensor $T - \frac{1}{3}g^{-1}Tr(T)$ has a vanishing trace in all reference frames.

Let us finally notice that, in the new complex notation, the Laplacian operator is written as $\nabla^2 = 4\partial_{x_+}\partial_{x_-} + \partial_z^2$. That can be easily verified by observing that the reality condition gives $\nabla^2 = k\partial_{x_+}\partial_{x_-} + \partial_z^2$ and that the value of k is fixed by the well known relation $\nabla^2 r^2 = 6$.

Going back to the solution of (B.1), let us notice that, until now, we have neglected rotations around axes other than z. If we consider for instance rotations around the y axis, since they change the z component of vectors, hence also of the angular momentum, they will change the generic solution $\psi_{E,m}(\boldsymbol{r})$ in a linear combination of solutions of the same kind $\psi_{E,m'}(\boldsymbol{r})$, corresponding to the same E; in particular a rotation of $180°$ changes the sign of m because it reverses the z axis. This means that, in the case of a central symmetry, energy levels are surely degenerate if $m \neq 0$.

Actually, what we are looking for are *degenerate multiplets* being also *irreducible* under rotations, meaning by that the minimal multiplets of solutions having the same energy E and transforming into each other under rotations in such a way that no sub-multiplet exists whose components do not mix with the remnant of the multiplet: the reason for doing so is that in this way we shall automatically characterize the solutions of the Schrödinger equation according to their transformations properties under the symmetry of the problem, finding also many features of them which are valid independently of the particular central field under consideration.

Being inspired by Taylor expansions, we will study wave functions having the following analytic structure

$$\psi_E^l(\boldsymbol{r}) = \mathcal{P}_l(\boldsymbol{r}) f_E(r) \; , \tag{B.4}$$

where $\mathcal{P}_l(\boldsymbol{r})$ is a generic homogeneous polynomial of degree l in the components of $\boldsymbol{r}$, i.e. in z and $x_\pm$. The reason for choosing homogeneous polynomials is that rotations act as homogeneous linear transformations on the components of $\boldsymbol{r}$ leaving r invariant, hence they transform the generic multiplet in (B.4) with a fixed choice for f_E into itself. Moreover, the multiplet described by (B.4) contains as special cases the functions of the type given in (B.2), for all values of m such that $-l \le m \le l$. Indeed two particular choices of $\mathcal{P}_l(\boldsymbol{r})$ are the functions x_+^l and x_-^l: they correspond to the maximum value of $|m|$. The fact that rotations around the y axis mix different values of m among themselves till they change their sign, shows that the set of functions described by (B.4) contain an irreducible multiple of at least $2l + 1$ elements.

On the other hand, let us check whether the multiplet made up of all possible homogeneous polynomials of degree l in x_+, x_-, z , which of course has $(l + 1)(l + 2)/2$ independent components, is irreducible or not. For $l = 1$ the number of such polynomials coincides with $2l + 1 = 3$, so that, according to what stated above, they are necessarily irreducible; in particular the three polynomials in this case can be identified with the components of the position vector $\boldsymbol{r}$, which are x_+, x_- and z, corresponding respectively to $m = 1, -1, 0$. However the same is not true already for $l = 2$: in this case $(l+1)(l+2)/2 = 6$ and a possible basis is given by the components of the symmetric tensor of rank 2 made up of two position vectors, $T_{i,j} = r_i r_j$, which has indeed 6 independent components. It is clear, however, that the combination $z^2 + x_+ x_- = r^2$ does not change under rotations, so that it does not mix with the other components: we must therefore eliminate it from the list of our polynomials, the remnant being constructed with a set of 5 independent polynomials, which must be irreducible since $2l + 1 = 5$ for $l = 2$, and can be chosen in correspondence with $m = -2, -1, 0, 1, 2$. In fact the invariant polynomial we have left out for $l = 2$ corresponds to the trace of the tensor introduced above, and the irreducible representation of dimension 5 is nothing but the corresponding traceless symmetric tensor.

Proceeding further to $l = 3$, we notice that the set of all possible polynomials has dimension $(l + 1)(l + 2)/2 = 10$. Among them an irreducible sub-multiplet of dimension 3 can be easily identified, corresponding to $r^2(x_+, x_-, z)$ (that being the trace of the rank 3 symmetric tensor $T_{i,j,k} = r_i r_j r_k$), plus a remnant of 7 independent polynomials which, since $2l + 1 = 7$ for $l = 3$, must necessarily be irreducible and can be chosen in correspondence with $m = -3, -2, \ldots, 3$. It should be clear that, with the aim of describing our solutions according to the general form proposed in (B.4), only the last irreducible multiplet, which contains the maximum possible value of m, i.e. $m = l$, is actually relevant, since the sub-multiplet of dimension 3 is nothing but a repetition of the case $l = 1$, as it is evident if we include the factor r^2 into the function $f_E(r)$.

It is of course not convenient and quite cumbersome to proceed as above for each single l separately. Let us therefore try to construct explicitly, for a generic l, the irreducible multiplet containing the polynomial with the maximum possible value of m, i.e. $\mathcal{P}_l(\boldsymbol{r}) = x_+^l$, which is surely not proportional to any invariant factor like r^2 (hence refers directly to the factorization given in (B.4)) and has at least dimension $2l + 1$. In order to do that, we notice that x_+^l is a solution of the equation $\nabla^2 x_+^l = 0$, i.e. it is a harmonic function. Since the Laplacian operator is clearly invariant under rotations, we must conclude that a generic rotation will transform x_+^l into a harmonic polynomial $\mathcal{P}_l(\boldsymbol{r})$, i.e. such that $\nabla^2 \mathcal{P}_l(\boldsymbol{r}) = 0$. That suggests to identify the irreducible multiplet we are looking for with the set of harmonic polynomials of order l. If the number of independent harmonic polynomials is $2l + 1$, the identification is demonstrated.

In order to verify that, let us explicitly construct a basis of independent homogeneous harmonic polynomials of degree l. Taking into account the form of the Laplacian, it can be easily checked that such basis can be chosen according to:

$$\mathcal{P}_l^m(\boldsymbol{r}) = x_+^m \sum_{k=0}^{[(l-m)/2]} \frac{(-\frac{1}{4})^k}{k!} \frac{(l-m)!m!}{(m+k)!(l-m-2k)!} (x_+ x_-)^k z^{l-m-2k} \quad (B.5)$$

for $m \geq 0$, and by:

$$\mathcal{P}_l^{-m}(\boldsymbol{r}) \equiv (\mathcal{P}_l^m(\boldsymbol{r}))^* \quad (B.6)$$

otherwise. Here $[(l-m)/2]$ stands for the integer part of $(l-m)/2$. We notice that this basis consists of $2l + 1$ homogeneous polynomials corresponding to all possible values of m such that $-l \leq m \leq l$. They change under rotations around the z axis by a phase factor $e^{im\phi}$, hence we conclude that, on the basis of previous considerations, the polynomials $\mathcal{P}_l^m(\boldsymbol{r})$ represent a basis for a multiplet which is irreducible under rotations.

At this point, since we are convinced that an irreducible multiplet of solutions of (B.1) can be put in the form: $f_{E,l}(r)\mathcal{P}_l^m(\boldsymbol{r})$, let us substitute this form into the mentioned equation. Let us consider at first the term involving the Laplacian operator. Noticing that $\mathcal{P}_l^m(\boldsymbol{r})$ is harmonic, that $\boldsymbol{\nabla} r = \boldsymbol{r}/r$ and that $\boldsymbol{r} \cdot \boldsymbol{\nabla} \mathcal{P}_l^m(\boldsymbol{r}) = l\mathcal{P}_l^m(\boldsymbol{r})$ since $\mathcal{P}_l^m(\boldsymbol{r})$ is a homogeneous polynomial of degree l, we can easily evaluate

$$\nabla^2 f_{E,l}(r)\mathcal{P}_l^m(\boldsymbol{r}) = \mathcal{P}_l^m(\boldsymbol{r})\nabla^2 f_{E,l}(r) + 2\boldsymbol{\nabla} f_{E,l}(r) \cdot \boldsymbol{\nabla} \mathcal{P}_l^m(\boldsymbol{r})$$

$$= \mathcal{P}_l^m(\boldsymbol{r})\boldsymbol{\nabla} \cdot \left(\frac{\boldsymbol{r}}{r} f_{E,l}'(r)\right) + 2\frac{f_{E,l}'(r)}{r}\boldsymbol{r} \cdot \boldsymbol{\nabla} \mathcal{P}_l^m(\boldsymbol{r})$$

$$= \mathcal{P}_l^m(\boldsymbol{r})\left[f"_{E,l}(r) + 2(l+1)\frac{f_{E,l}'(r)}{r}\right], \quad (B.7)$$

hence, leaving the factor $\mathcal{P}_l^m(\boldsymbol{r})$ out, the stationary Schrödinger equation becomes:

$$-\frac{\hbar^2}{2m}\left[f''_{E,l}(r) + 2(l+1)\frac{f'_{E,l}(r)}{r}\right] + V(r)f_{E,l}(r) = Ef_{E,l}(r)\,. \qquad (B.8)$$

If we make the substitution $f_{E,l} \equiv \chi_{E,l}/r^{l+1}$ and leave a factor $1/r^{l+1}$ out, we obtain:

$$-\frac{\hbar^2}{2m}\chi''_{E,l}(r) + \left[\frac{\hbar^2 l(l+1)}{2mr^2} + V(r)\right]\chi_{E,l}(r) = E\chi_{E,l}(r)\,, \qquad (B.9)$$

which is completely analogous to the one dimensional stationary Schrödinger equation in presence of the potential energy obtained by adding the term $\hbar^2 l(l+1)/(2mr^2)$ to $V(r)$. Recalling the *centrifugal* potential energy $L^2/2m$ appearing in the classical mechanics analysis of the motion in a central field, we can finally conclude that the wave function:

$$\psi_{E,l,m}(\boldsymbol{r}) = \frac{\chi_l(r)}{r^{l+1}}\mathcal{P}_l^m(\boldsymbol{r}) \equiv \frac{\chi_l(r)}{r}\Pi_l^m(\theta\,,\varphi) \qquad (B.10)$$

corresponds to a state with squared angular momentum $L^2 = \hbar^2 l(l+1)$ and $L_z = \hbar m$. In Equation (B.10) we have put into evidence that $\mathcal{P}_l^m(\boldsymbol{r})/r^l \equiv \Pi_l^m$ is a homogeneous function of degree zero in r, hence it only depends on polar angles.

In current textbooks the functions $\Pi_l^0(\theta)$ are indicated by $P_l(\cos\theta)$ and are called *Legendre polynomials* in the argument $\cos\theta$. It also common to make use of *spherical harmonics*:

$$Y_{l,m}(\theta,\varphi) \equiv \frac{(-1)^m}{m!}\sqrt{\frac{(2l+1)(l+m)!}{4\pi(l-m)!}}\,\Pi_l^m(\theta\,,\varphi) \quad \text{for} \quad m \ge 0\,,$$

$$Y_{l,-m}(\theta,\varphi) = (-1)^m Y^*_{l,m}(\theta,\varphi)\,, \qquad (B.11)$$

which satisfy the normalization condition:

$$\int_0^{2\pi} d\varphi \int_{-1}^{1} d\cos\theta\, Y^*_{l,m}(\theta,\varphi)Y_{l',m'}(\theta,\varphi) = \delta_{l,l'}\delta_{m,m'}\,. \qquad (B.12)$$

Spherical harmonics can be used in place of the functions Π_l^m in (B.10), furnishing the expressions which are commonly used for the wave functions in spherical coordinates; these wave functions can then be easily normalized on the basis of (B.12). Legendre polynomials satisfy instead the following relation

$$\int_{-1}^{1} dx\, P_l(x)P_{l'}(x) = \frac{2\delta_{l,l'}}{2l+1}\,, \qquad P_l(1) = 1\,. \qquad (B.13)$$

The functions $\mathcal{P}_l^m(\boldsymbol{r})$ introduced above are instead normalized according to the condition $\mathcal{P}_l^m(\boldsymbol{r}) \to_{\theta \to 0} x_+^m z^{l-m}$ for $m \geq 0$. Starting from $l = 0$ we find:

$$\mathcal{P}_0^0 = 1 \;, \qquad \mathcal{P}_1^0 = z \;, \qquad \mathcal{P}_1^{\pm 1} = x_\pm \;,$$

$$\mathcal{P}_2^0 = 3z^2 - r^2 \;, \qquad \mathcal{P}_2^{\pm 1} = x_\pm z \;, \qquad \mathcal{P}_2^{\pm 2} = (x_\pm)^2 \;. \qquad (\text{B}.14)$$

Therefore we have better specified what already said above: the functions $\mathcal{P}_1^m$ coincide with the complex components of the position vector $\boldsymbol{r}$, i.e. $r_\pm = x_\pm$, $r_3 = z$, while the functions $\mathcal{P}_2^m$ are proportional to the components of the traceless symmetric tensor

$$T_{ij}(\boldsymbol{r}) \equiv r_i r_j - \frac{1}{3} g_{ij}^{-1} r^2 \;. \qquad (\text{B}.15)$$

Indeed, it is evident that:

$$\mathcal{P}_2^0 = 3T_{33} \;, \qquad \mathcal{P}_2^{\pm} = T_{3\pm} \;, \qquad \mathcal{P}_2^{\pm 2} = T_{\pm\pm} \;. \qquad (\text{B}.16)$$

The relation between the wave functions of systems having central symmetry and traceless symmetric tensors, which generalizes what we have explicitly seen for $l = 2$, allows to easily understand how to combine the angular momenta of different components of the same system, a typical example being an atom emitting electromagnetic radiation, whose angular momentum is strictly related to the emission multipolarity. It is indeed quite easy to understand how different traceless symmetric tensors can combine to form new tensors of the same kind, through the direct product $U_{i_1,..,i_n} = T_{i_1,..,i_k} V_{i_{k+1},..,i_n}$ and the repeated use ot the trace operation[1], which gives $W_{i_1,..,i_{n-2}} = \sum_{jl} U_{i_1,..,i_{n-2},j,l}\, g^{jl}$. Being the original tensor already traceless, it is evident that non-vanishing traces can only be taken among pairs of indices not belonging to the same tensor; therefore, starting from two tensors of rank l and l', we can construct new traceless symmetric tensors of rank in the range between $l + l'$ and $|l - l'|$. The strict relation between the rank of the wave function, considered as a tensor, and angular momentum, let us immediately understand that angular momenta combine like the ranks of the corresponding tensors, hence the resulting angular momentum can only take values between $l + l'$ and $|l - l'|$, which by the way is also the possible range of lengths for a vector sum of two vectors of length l and l'. In an electric quadrupole transition, where the radiation enters through a traceless symmetric tensor of rank 2, hence $l' = 2$, the final angular momentum of the

---

[1] A further way of combining T and V exploits the completely antisymmetric invariant rank 3 tensor $\epsilon_{i,j,k}$ which is usually normalized according to $\epsilon_{x,y,z} = 1$. One takes the double trace involving ϵ, T and V. This results in a new tensor of rank $k + n - 1$ which can be symmetrized and further reduced separating its traces.

emitting atom may take values in the range between $l + 2$ and $|l - 2|$, where l is the original atomic angular momentum.

Let us now consider the solution of radial equations (B.8) and (B.9). In the case of piecewise constant potentials $V(r)$, similar to those discussed in the one dimensional case, the analysis strategy does not change and one has only to pay special attention to the additional constraint that the wave function must vanish in $r = 0$, otherwise the related probability density would be divergent. In particular in the S wave case (that being the usual way of indicating the case $l = 0$) Equation (B.9) is exactly equal to the one dimensional Schrödinger equation, therefore the solution can be obtained as a continuous linear combination of functions like $\sin(\sqrt{2m(E - V)}r/\hbar)$ and $\cos(\sqrt{2m(E - V)}r/\hbar)$ for $E > V$ or $\sinh(\sqrt{2m(E - V)}r/\hbar)$ and $\cosh(\sqrt{2m(E - V)}r/\hbar)$ in the opposite case. For $l > 0$ sine and cosine (hyperbolic or not) must be replaced by new special functions (which are called *spherical Bessel function*), which can be explicitly constructed by writing (B.9) like

$$\chi''_{E,l}(r) + \left[\pm k^2 - \frac{l(l + 1)}{r^2}\right] \chi_{E,l}(r) = 0 , \qquad (B.17)$$

where we have set $k^2 = 2m|E - V|/\hbar^2$. The validity of the following recursive equation can be directly checked:

$$k \,\chi_{E,l+1}(r) = \frac{l + 1}{r}\chi_{E,l}(r) - \chi'_{E,l}(r) . \qquad (B.18)$$

If for instance we want to study the possible bound states in P wave (i.e. $l = 1$) in the potential well: $V(r) = -V_0$ for $r < R$ and $V = 0$ otherwise, setting the energy to $-B$ and defining $\kappa = \sqrt{2mB/\hbar^2}$, we must connect the internal solution $\sin(qr)/qr - \cos(qr)$, which is regular in $r = 0$, with the external solution which vanishes as $r \to \infty$, i.e. $e^{-\kappa r}(1/\kappa r - 1)$.

Another case of great interest is obviously the study of bound states in a Coulomb potential, which permits an analysis of the energy levels of the hydrogen atom. To that aim let us consider the motion of a particle of mass m in a central potential $V(r) = -e^2/(4\pi\epsilon_0 r)$, where ϵ_0 is the vacuum dielectric costant and e is (minus) the charge of the (electron) proton in MKS units; m is actually the reduced mass in the case of the proton-electron system, $m = m_e m_p/(m_e + m_p)$, which is practically equal to the electron mass within a good approximation. In this case it is convenient to start from (B.8), which we rewrite as

$$f"_{B,l}(r) + 2(l + 1)\frac{f'_{B,l}(r)}{r} + \frac{2me^2}{4\pi\epsilon_0\hbar^2 r}f_{B,l}(r) = \frac{2mB}{\hbar^2}f_{B,l}(r) , \qquad (B.19)$$

where $B \equiv -E$ is the binding energy. Before proceeding further let us perform a change of variables in order to work with adimensional quantities: we introduce Bohr's radius $a_0 = 4\pi\epsilon_0\hbar^2/(me^2) \simeq 0.52 \; 10^{-10}$ m and Rydberg's energy constant $E_R \equiv hR \equiv me^4/(2\hbar^2(4\pi\epsilon_0)^2) \simeq 13.6$ eV, which are the typical length and energy scales which can be constructed in terms of the physical constants involved in the problem. Equation (B.19) can be rewritten in terms of the adimensional radial variable $\rho \equiv r/a_0$ and of the adimensional binding energy $B/E_R \equiv \lambda^2$, with $\lambda \geq 0$, as follows:

$$f"_{\lambda,l}(\rho) + 2(l+1)\frac{f'_{\lambda,l}(\rho)}{\rho} + \frac{2}{\rho}f_{\lambda,l}(\rho) = \lambda^2 f_{\lambda,l}(\rho) \,. \tag{B.20}$$

Let us first consider the asymptotic behavior of the solution as $\rho \to \infty$: in this limit the second and the third term on the left hand side can be neglected, so that the solution of (B.20) is asymptotically also solution of $f"_{\lambda,l}(\rho) = \lambda^2 f_{\lambda,l}(\rho)$, i.e. $f_{\lambda,l}(\rho) \sim e^{\pm\lambda\rho}$ for $\rho \gg 1$. The asymptotically divergent behavior must obviously be rejected if we are looking for a solution corresponding to a normalizable bound state. We shall therefore write our solution in the form $f_{\lambda,l}(\rho) = h_{\lambda,l}(\rho)e^{-\lambda\rho}$, where $h_{\lambda,l}(\rho)$ must be a sufficiently well behaved function as $\rho \to \infty$. The differential equation satisfied by $h_{\lambda,l}(\rho)$ easily follows from (B.20):

$$h''_{\lambda,l} + \left(\frac{2(l+1)}{\rho} - 2\lambda\right)h'_{\lambda,l} + \frac{2}{\rho}\left(1 - \lambda(l+1)\right)h_{\lambda,l} = 0 \,. \tag{B.21}$$

We shall write $h_{\lambda,l}(\rho)$ as a power series in ρ, thus finding a recursion relation for its coefficients, and impose that the series stops at some finite order so as to keep the asymptotic behaviour of $f_{\lambda,l}(\rho)$ as $\rho \to \infty$ unchanged.

In order to understand what is the first term ρ^s of the series that we must take into account, let us consider the behavior of (B.21) as $\rho \to 0$. In this case $h_{\lambda,l} \sim \rho^s$ and it can be easily checked that (B.21) is satisfied at the leading order in ρ only if $s(s-1) = -2s(l+1)$, whose solutions are $s = 0$ and $s = -2l - 1$. Last possibility must be rejected, otherwise the probability density related to our solution would be divergent in the origin. Hence we write:

$$h_{\lambda,l}(\rho) = c_0 + c_1\rho + c_2\rho^2 + \ldots + c_h\rho^h + \ldots = \sum_{h=0}^{\infty} c_h\rho^h \,, \tag{B.22}$$

with $c_0 \neq 0$. Inserting last expression in (B.21), we obtain the following recurrence relation for the coefficients c_h:

$$c_{h+1} = 2\frac{\lambda(h+l+1) - 1}{(h+1)(h+2(l+1))}c_h \tag{B.23}$$

which, apart from an overall normalization constant fixing the starting coefficient c_0, completely determines our solution in terms of l and λ. However if the recurrence relation never stops, it becomes asymptotically (i.e. for large h):

$$c_{h+1} \simeq \frac{2\lambda}{h} c_h$$

which can be easily checked to be the same relation relating the coefficients in the Taylor expansion of $\exp(2\lambda\rho)$. Therefore, if the series does not stop, the asymptotic behavior of $f_{\lambda,l}(\rho)$ will be corrupted, bringing in fact back the unwanted divergent behavior $f_{\lambda,l}(\rho) \sim e^{\lambda\rho}$. The series stops if and only if the coefficient on the right hand side of (B.23) vanishes for some given value $h = k \geq 0$, hence

$$\lambda(k+l+1) - 1 = 0 \quad \Rightarrow \quad \lambda = \frac{1}{k+l+1} . \tag{B.24}$$

In this case $h_{\lambda,l}(\rho)$ is simply a polynomial of degree k in ρ, which is completely determined (neglecting an overall normalization) as a function of l and k: these polynomials belong to a well know class of special functions and are usually called Laguerre's associated polynomials. We have therefore found that, for a given value of l, the admissible solutions with negative energy, i.e. the hydrogen bound states, can be enumerated according to a non-negative integer k and the energy levels are quantized according to (B.24).

If we replace k by a new integer and strictly positive quantum number n given by:

$$n = k+l+1 = \lambda^{-1} , \tag{B.25}$$

which is usually called the *principal quantum number*, then the energy levels of the hydrogen atom, according to (B.24) and to the definition of λ, are given by

$$E_n = -\frac{E_R}{n^2} = -\frac{me^4}{8\epsilon_0^2 h^2 n^2} ,$$

in perfect agreement with the Balmer–Rydberg series for line spectra and with the qualitative result obtained in Section 2.2 using Bohr's quantization rule.

It is important to notice that, in the general case of a motion in a central field, energy levels related to different values of the angular momentum l are expected to be different, and they are indeed related to the solutions of different equations of the form given in (B.8). Stated otherwise, the only expected degeneracy is that related to the rotational symmetry of the problem, leading to degenerate wave function multiplets of dimension $2l+1$, as discussed above. However in the hydrogen atom case we have found a quite different result: according to (B.25), for a fixed value of the integer $n > 0$, there will be n different multiplets, corresponding to $l = 0, 1, \ldots, (n-1)$, having the same energy. The degeneracy is therefore

$$\sum_{l=0}^{n-1} (2l+1) = n^2$$

instead of $2l+1$. Unexpected additional degeneracies like this one are usually called "accidental", even if in this case it is not so accidental. Indeed the problem of the motion in a Coulomb (or gravitational) field has a larger symmetry than simply the rotational one. We will not go into details, but just remind the reader of a particular integral of motion which is only present, among all possible central potentials, in the case of the Coulomb (gravitational) field: that is Lenz's vector, which completely fixes the orientation of classical orbits. Another central potential leading to a similar "accidental" degeneracy is that corresponding to a three-dimensional isotropic harmonic oscillator. Actually, the Coulomb potential and the harmonic oscillator are joined in Classical Mechanics by Bertrand's theorem, which states that they are the only central potentials whose classical orbits are always closed.

Let us finish by giving the explicit form of the hydrogen wave functions in a few cases. Writing them in a form similar to that given in (B.10), and in particular as

$$\psi_{n,l,m}(r, \theta, \phi) = R_{n,l}(r) Y_{l,m}(\theta, \varphi),$$

we have

$$R_{1,0}(r) = 2(a_0)^{-3/2} \exp(-r/a_0),$$

$$R_{2,0}(r) = 2(2a_0)^{-3/2} \left(1 - \frac{r}{2a_0}\right) \exp(-r/2a_0),$$

$$R_{2,1}(r) = (2a_0)^{-3/2} \frac{1}{\sqrt{3}} \frac{r}{a_0} \exp(-r/2a_0).$$

C

Thermodynamics and Entropy

We have shown by some explicit examples how the equilibrium distribution of a given system can be found once its energy levels are known. That has allowed us to compute the mean equilibrium energy by identifying the Lagrange multiplier β with $1/kT$. However, it should be clear that the complete reconstruction of the thermodynamics properties of systems in equilibrium requires some further steps and more information.

Nevertheless, in the case of Einstein's crystal in the limit of a rigid lattice, the thermodynamical analysis is quite simple. Indeed the model describes a system whose only energy exchanges with the external environment happen through heat transfer. That means that the exchanged heat is a function of the state of the system which does not differ but for an additive constant from the mean energy, i.e. from the internal energy $U(T)$. However also in this simple case we can introduce the concept of entropy S, starting from the differential equation $dS = dU/T$, which making use of (3.23) gives:

$$dS = \frac{C(T)}{T}dT = \frac{3\hbar^2\omega^2}{kT^3}\frac{e^{\hbar\omega/kT}}{\left(e^{\hbar\omega/kT} - 1\right)^2}dT = -3k\hbar^2\omega^2\frac{e^{\beta\hbar\omega}}{\left(e^{\beta\hbar\omega} - 1\right)^2}\beta d\beta$$

$$= d\left[3k\left(\frac{\beta\hbar\omega}{e^{\beta\hbar\omega} - 1} - \ln\left(1 - e^{-\beta\hbar\omega}\right)\right)\right]. \tag{C.1}$$

Hence, if we choose $S(0) = 0$ as the initial condition for $S(T)$, we can easily write the entropy of Einstein's crystal:

$$S(T) = \frac{3\hbar\omega}{T}\frac{1}{e^{\hbar\omega/kT} - 1} - 3k\ln\left(1 - e^{-\hbar\omega/kT}\right), \tag{C.2}$$

showing in particular that at high temperatures $S(T)$ grows like $3k\ln T$.

Apart from this result, equation (C.2) is particularly interesting since it can be simply interpreted in terms of statistical equilibrium distributions. Indeed, recalling that the probability of the generic state of the system, which is identified with the vector $\boldsymbol{n}$, is given by:

$$p_n = e^{-\beta\hbar\omega(n_x+n_y+n_z)} \left(1 - e^{-\beta\hbar\omega}\right)^3 ,$$

we can compute the following expression:

$$-k \sum_{n_x,n_y,n_z=0}^{\infty} p_n \ln p_n$$

$$= k \left(1 - e^{-\beta\hbar\omega}\right)^3 \sum_{n_x,n_y,n_z=0}^{\infty} e^{-\beta\hbar\omega(n_x+n_y+n_z)}$$

$$\left[\beta\hbar\omega \left(n_x + n_y + n_z\right) - 3\ln\left(1 - e^{-\beta\hbar\omega}\right)\right]$$

$$= -3k\ln\left(1 - e^{-\beta\hbar\omega}\right) \left(\left(1 - e^{-\beta\hbar\omega}\right) \sum_{n=0}^{\infty} e^{-\beta\hbar\omega n}\right)^3$$

$$+3k\,\beta\,\hbar\,\omega \left(1 - e^{-\beta\hbar\omega}\right)^3 \left(\sum_{n=0}^{\infty} e^{-\beta\hbar\omega n}\right)^2 \sum_{m=0}^{\infty} m\, e^{-\beta\hbar\omega m} . \qquad \text{(C.3)}$$

Finally, making use of the expression for the geometric series $\sum_{n=0}^{\infty} x^n = 1/(1-x)$ if $|x| < 1$, hence

$$\sum_{n=1}^{\infty} n x^n = x \frac{d}{dx}\frac{1}{1-x} = \frac{x}{(1-x)^2} ,$$

we easily find again the expression in (C.2).

One of the most important consequences of this result is the probabilistic interpretation of entropy following from the equation

$$S = -k \sum_{\alpha} p_\alpha \ln p_\alpha , \qquad \text{(C.4)}$$

which has a general validity and is also particularly interesting for its simplicity. Indeed, let us consider an isolated system and assume that its accessible states be equally probable, so that p_α is constant and equal to the inverse of the number of states. Indicating that number by Ω, it easily follows that $S = k\ln\Omega$, hence entropy measures the number of accessible states.

As an example, let us apply (C.4) to the three level system studied in Section 3.3. In this case entropy can be simply expressed in terms of the parameter $z = e^{-\beta\epsilon}$:

$$S = k\ln(1 + z + z^2) - \frac{k}{1 + z + z^2}(z + 2z^2)\ln z ,$$

which has a maximum for $z = 1$, i.e. at the border of the range corresponding to possible thermal equilibrium distributions.

In order to discuss the generality of (C.4) we must consider the general case in which the system can exchange work as well as heat with the external environment. For instance Einstein's model could be made more realistic by assuming that, in the relevant range of pressures, the frequencies of oscillators depend on their density according to $\omega = \alpha \left(N/V\right)^{\gamma}$, where typically $\gamma \sim 2$. In these conditions the crystal exchanges also work with the external environment and the pressure can be easily computed by using (3.32):

$$P = \sum_{n_x,n_y,n_z=0}^{\infty} p_n \frac{\gamma}{V} E_n = \frac{\gamma}{V} \frac{U}{V}, \tag{C.5}$$

thus giving the equation of state for the crystal.

Going back to entropy, let us compute, in the most general case, the heat exchanged when the parameters β and V undergo infinitesimal variations. From (3.32) we get:

$$dU + PdV = \frac{\partial U}{\partial \beta} d\beta + \frac{\partial U}{\partial V} dV + \frac{1}{\beta} \frac{\partial \ln Z}{\partial V} dV, \tag{C.6}$$

hence, making use of (3.15) we obtain the following infinitesimal heat transfer:

$$-\frac{\partial^2 \ln Z}{\partial \beta^2} d\beta - \frac{\partial^2 \ln Z}{\partial \beta \partial V} dV + \frac{1}{\beta} \frac{\partial \ln Z}{\partial V} dV. \tag{C.7}$$

Last expression can be put into the definition of entropy, thus giving:

$$\begin{aligned} dS &= k\beta \left[-\frac{\partial^2 \ln Z}{\partial \beta^2} d\beta - \frac{\partial^2 \ln Z}{\partial \beta \partial V} dV + \frac{1}{\beta} \frac{\partial \ln Z}{\partial V} dV \right] \\ &= k \left[\frac{\partial}{\partial \beta} \left(\ln Z - \beta \frac{\partial \ln Z}{\partial \beta} \right) d\beta + \frac{\partial}{\partial V} \left(\ln Z - \beta \frac{\partial \ln Z}{\partial \beta} \right) dV \right] \\ &= k\, d \left(\ln Z - \beta \frac{\partial \ln Z}{\partial \beta} \right). \end{aligned} \tag{C.8}$$

On the other hand it can be easily verified, using again (3.15), that:

$$-k \sum_{\alpha} p_\alpha \ln p_\alpha = k \sum_{\alpha} p_\alpha \left(\beta E_\alpha + \ln Z \right)$$

$$= k \left(\beta U + \ln Z \right) = k \left(\ln Z - \beta \frac{\partial \ln Z}{\partial \beta} \right) \equiv S. \tag{C.9}$$

Last equation confirms that the probabilistic interpretation of entropy has a general validity and also gives an expression of the partition function in terms of thermodynamical potentials. Indeed, the relation $S = k \left(\beta U + \ln Z \right)$ is equivalent to $U = TS - \ln Z/\beta$, hence to $\ln Z = -\beta(U - TS)$. Therefore we can conclude that the logarithm of the partition function equals minus β times the free energy.

Index

Printed in June 2007